Data-Driven Intelligence in Wireless Networks

This book highlights the importance of data-driven techniques to solve wireless communication problems. It presents a number of problems (e.g., related to performance, security, and social networking), and provides solutions using various data-driven techniques, including machine learning, deep learning, federated learning, and artificial intelligence.

This book details wireless communication problems that can be solved by data-driven solutions. It presents a generalized approach toward solving problems using specific data-driven techniques. The book also develops a taxonomy of problems according to the type of solution presented and includes several case studies that examine data-driven solutions for issues such as quality of service (QoS) in heterogeneous wireless networks, 5G/6G networks, and security in wireless networks.

The target audience of this book includes professionals, researchers, professors, and students working in the field of networking, communications, machine learning, and related fields.

Data-Driven Intelligence in Wireless Networks

Concepts, Solutions, and Applications

Edited by
Muhammad Khalil Afzal,
Muhammad Ateeq, and
Sung Won Kim

CRC Press
Taylor & Francis Group
Boca Raton New York London

CRC Press is an imprint of the
Taylor & Francis Group, an **Informa** business

Designed cover image: iStock – Smart City and Internet of Things, Various Communication Devices Stock Photo

MATLAB® is a trademark of The MathWorks, Inc. and is used with permission. The MathWorks does not warrant the accuracy of the text or exercises in this book. This book's use or discussion of MATLAB® software or related products does not constitute endorsement or sponsorship by The MathWorks of a particular pedagogical approach or particular use of the MATLAB® software.

First edition published 2023
by CRC Press
6000 Broken Sound Parkway NW, Suite 300, Boca Raton, FL 33487-2742

and by CRC Press
4 Park Square, Milton Park, Abingdon, Oxon, OX14 4RN

CRC Press is an imprint of Taylor & Francis Group, LLC

© 2023 selection and editorial matter, Muhammad Khalil Afzal, Muhammad Ateeq, and Sung Won Kim; individual chapters, the contributors

Library of Congress Cataloging-in-Publication Data

Names: Afzal, Muhammad Khalil, editor. | Ateeq, Muhammad, editor. | Kim, Sung Won, editor.
Title: Data-driven intelligence in wireless networks : concepts, solutions, and applications / edited by Muhammad Khalil Afzal, Muhammad Ateeq, and Sung Won Kim.
Description: First edition. | Boca Raton : CRC Press, 2023. | Includes bibliographical references. |
Identifiers: LCCN 2022041502 (print) | LCCN 2022041503 (ebook) | ISBN 9781032100371 (hardback) | ISBN 9781032107738 (paperback) | ISBN 9781003216971 (ebook)
Subjects: LCSH: Wireless communication systems--Design and construction--Data processing. | Decision making. | Big data. | Artificial intelligence.
Classification: LCC TK5103.2 .D3157 2023 (print) | LCC TK5103.2 (ebook) | DDC 621.3820285/63--dc23/eng/20221115
LC record available at https://lccn.loc.gov/2022041502
LC ebook record available at https://lccn.loc.gov/2022041503

ISBN: 978-1-032-10037-1 (hbk)
ISBN: 978-1-032-10773-8 (pbk)
ISBN: 978-1-003-21697-1 (ebk)

DOI: 10.1201/9781003216971

Typeset in Times
by KnowledgeWorks Global Ltd.

Contents

PART I   Data-Driven Wireless Networks: Design and Applications

PART II   Data-Driven Techniques and Security Issues in Wireless Networks

PART III   Advanced Topics in Data-Driven Intelligence for Wireless Networks

Preface

An evolving concept called data-driven intelligence is a model for a new viewpoint on gathering vision from a vast pool of data. Data-driven techniques put robust importance on the large dataset to solve a specific problem. There are more than 370 million Internet users worldwide. The number of unique mobile users is almost 5 billion, with the total number of mobile connections exceeding 8 billion. This indicates that wireless communication is a prevalent field. Wireless networks can show random interactions between algorithms from multiple protocol layers, interactions between multiple devices, and hardware-specific effects. Different wireless technologies including mobile cellular, fixed line, Wi-Fi, and others are widely used for diverse communication purposes. It is estimated that the majority of users access the Internet via mobile devices. Internet traffic is dominated by multimedia content, and its proportion is ever increasing. This increases the quality of service/quality of experience requirements. Optimizing multiple, often conflicting goals according to different application requirements is of fundamental importance in this context. With the spreading of the Internet and broadening of its capacity, data is available in abundance. Moreover, more wireless data can be collected from wireless testbed facilities. Many fields benefit from data to optimize decisions. Many data sets related to the performance and security of different wireless networks, i.e., wireless sensor networks and the Internet of Things, are publicly available. There has been a growing trend to benefit from data-driven techniques like machine learning to improve decision-making, management, performance, and security issues in wireless networks.

This edited book aims to deliver knowledge that can highlight the importance of data-driven techniques to solve wireless communication problems. As a next step, the solution to those problems, using various data-driven techniques, primarily from machine learning (supervised, unsupervised, and reinforcement), deep learning, federated learning, and artificial intelligence, are presented.

The chapters of this book are authored by several international researchers. This book is composed of **10** chapters that can be read based on the interest of the reader without having to read the entire book. These chapters were carefully selected after a rigorous review. This book is an excellent reference for computer scientists, researchers, and developers, who wish to contribute to the domain of data-driven intelligence in the field of wireless networks and related areas. We tried to include sufficient details and provide the necessary background information in each chapter to help the readers easily understand the content. We hope readers will enjoy this book and hope it will help graduate students who are interested in working on the domain of data-driven intelligence in their research.

Acknowledgments

We would like to express our gratitude to everyone who participated in this book and made this book a reality. We would especially like to acknowledge the hard work of the authors.

We would also like to acknowledge the efforts of the reviewers, whose valuable comments enabled us to select these chapters out of the many we received and whose help improved the overall quality of the chapters presented in this book. Special thanks to Dr. Wazir Zada Khan, University of Wah, Pakistan; Dr. Yousaf Bin Zikria, Yeungnam University, South Korea; Muhammad Islam, Swinburne University of Technology, Australia; Dr. Zulqarnain, Yeungnam University, South Korea; Dr. Salman Saadat, Military College, Oman; Dr. Zeeshan Kaleem, COMSATS University Islamabad, Wah Campus; Dr. Latif Ullah Khan, Kyung Hee University, South Korea; Naqqash Dilshad, Sejong University, South Korea; and Ahsan Saleem, Concordia University, Canada.

Lastly, we are very grateful to the editorial team at Taylor & Francis for their support throughout the publishing of this book. Special Thanks to our editor Marc Gutierrez, for his great support and encouragement. We would also like to thank Sarahjayne Smith from Taylor & Francis, and Deepanshu from KnowledgeWorks Global Ltd., for managing the production process of this book.

Dr. Muhammad Khalil Afzal
Dr. Muhammad Ateeq
Dr. Sung Won Kim

Editor Biographies

Muhammad Khalil Afzal (SM'16) received his MCS and MS degrees in Computer Science from COMSATS Institute of Information Technology, Wah Campus, Pakistan in 2004 and 2007, respectively, and his Ph.D. from the Department of Information and Communication Engineering, Yeungnam University, South Korea, in December 2014. He has served as a lecturer from January 2008 to November 2009 in Bahauddin Zakariya University, Pakistan, and from December 2009 to June 2011 in King Khalid University Abha, Saudi Arabia. Currently, he is working as an Assistant Professor in the Department of Computer Science at COMSATS, Wah Cantt Pakistan. He has served as a Guest Editor of IEEE Internet of Things Journal, Elsevier Computer Communication, IEEE Communication Magazine, Transactions for Emerging Telecommunications Technologies (ETT), Future Generation Computer Systems (Elsevier), IEEE ACCESS, Journal of Ambient Intelligence and Humanized Computing (Springer), MDPI Sensors Journal and reviewer for IEEE ACCESS, Computers and Electrical Engineering (Elsevier), Journal of Network and Computer Applications (Elsevier), FGCS, and IEEE transaction on Vehicular Technology. He is the recipient of the best paper awards of the ACM 7th International Conference on Computing Communication and Networking Technologies (2016) and the International Conference on Green and Human Information Technology (ICGHIT-2018) and received a fully funded scholarship for his Masters and Ph.D. He is the recipient of a research project funded by the Higher Education Commission (HEC), Pakistan, and National Grassroots ICT Research Initiative, Ignite. His research interests include wireless sensor networks, ad hoc networks, data-driven intelligence in wireless networks, Smart Cities, 5G, and IoT.

Muhammad Ateeq received his bachelor's degree from Bahauddin Zakariya University at Multan in 2005, and a master's degree in Computer Science from the COMSATS Institute of Information Technology, Wah Cantonment, in 2007. He did his Ph.D. in Computer Science from the Department of Computer Science, COMSATS University Islamabad, Wah Campus Pakistan in June 2021. He has been in academia for the last 11 years. He is currently an Assistant Professor of Computer Science with The Islamia University of Bahawalpur. His research interest includes using data-driven techniques to improve the quality of service in wireless communication.

 **Sung Won Kim** received both a B.S. and M.S. degree from the Department of Control and Instrumentation Engineering, Seoul National University, South Korea, in 1990 and 1992, respectively, and a Ph.D. from the School of Electrical Engineering and Computer Sciences, Seoul National University, in 2002. From 1992 to 2001, he was a researcher with the Research and Development Center, LG Electronics, South Korea. From 2001 to 2003, he was a researcher with the Research and Development Center, AL Tech, South Korea. From 2003 to 2005, he was a postdoctoral researcher with the Department of Electrical and Computer Engineering, University of Florida, Gainesville, FL, USA. In 2005, he joined the Department of Information and Communication Engineering, Yeungnam University, Gyeongsan, South Korea, where he is currently a professor. His research interests include resource management, wireless networks, mobile networks, performance evaluation, and embedded systems.

Contributors

Muhammad Khalil Afzal
Department of Computer Science
COMSATS University Islamabad,
    Wah Campus
Pakistan

Lina Elmoiz Alatabani
Department of Data Communication
    and Network Engineering
The Future University
Sudan

Elmustafa Sayed Ali
Department of Electrical Engineering
Red Sea University
Sudan

Rashid Ali
Department of Software Device
    Engineering
Sejong University
Seoul, Republic of Korea

Rashid Amin
Department of Computer Science
University of Engineering of Technology
Taxila, Pakistan
University of Chakwal
Chakwal, Pakistan

Ahmad Arsalan
Department of Computer Science and
    Information Technology
University of Central Punjab
Pakistan

Muhammad Ateeq
Department of Data Science, Faculty
    of Computing
The Islamia University of Bahawalpur
Pakistan

Saima Bibi
Department of Computer Science
University of Engineering and
    Technology
Taxlia, Pakistan

Abdellah Chehri
Department of Mathematics and
    Computer Science
Royal Military College of Canada
Canada

Mudassar Hussain
Department of Computer Science
University of Sialkot
Pakistan

Hyung Seok Kim
College of Software Convergence
Sejong University
Seoul, Republic of Korea

Zhihan Lv
Department of Game Design
Faculty of Arts
Uppsala University
Sweden

El Mehdi Ouafiq
Hassania School of Public Works
Casablanca, Morocco

Rana Asif Rehman
Department of Computer Science
National University of Computer and
    Emerging Sciences, Lahore Campus
Pakistan

Rachid Saadane
Hassania School of Public Works
Casablanca, Morocco

Salman Saadat
Systems Engineering Department
Military Technological College
Oman

Ayesha Sabir
Department of Computer Science
University of Engineering and
    Technology
Taxila, Pakistan

Mamoon M. Saeed
University of Modern Sciences
Yemen

Rashid A. Saeed
Taif University
Saudi Arabia

Tariq Umer
Department of Computer Science
COMSATS University, Lahore Campus
Pakistan

Part I

Data-Driven Wireless Networks: Design and Applications

1 Data-Driven Wireless Networks

A Perspective

Muhammad Ateeq
Department of Data Science, Faculty of Computing,
The Islamia University of Bahawalpur, Pakistan

Muhammad Khalil Afzal
Department of Computer Science, COMSATS
University Islamabad, Wah Campus, Pakistan

CONTENTS

1.1 EVOLUTION OF WIRELESS NETWORKS

In modern communication systems, wireless has become a prevailing medium in establishing the Internet of Things (IoT) that uses all sorts of communication networks including WiFi, wireless sensor networks, and cellular networks. In our physical environment, we are surrounded by an enormous amount of information that we sense and to which we appropriately react. With the progression of the world and fast-growing technology, environments are intended to be smart where both sensing and decision-making are autonomous, fast, and robust. Such ubiquitous sensing requires support and facilitation to put the sensed data to use. There is an established

DOI: 10.1201/9781003216971-2

need for resilient, adaptive, and futuristic communication infrastructure. To serve this purpose, wireless sensor networks (WSNs) [1] have been around for the past two decades. WSNs consist of resource-constrained nodes capable of sensing and communicating over low-power radio interfaces. The number of nodes may range from in the 10s to the 1000s, depending on the application domain and deployment scenario. These nodes collect large-scale data and relay it to a base station that carries out the relevant processing tasks [2].

To be specific, we shall refer more to WSNs as an example of wireless networks through this chapter.

1.1.1 WSN Applications

WSNs target broad applications, primarily categorized into monitoring and tracking [3]. Some important and popular domains of interest include environment, industry and agriculture, infrastructure, military, body area, etc. In the context of industrial WSNs (IWSNs), the International Society of Automation (ISA) distributes industrial systems into the following six classes [4]:

- Safety/emergency systems
- Closed-loop regulatory control systems
- Closed-loop supervisory systems
- Open-loop control systems
- Alerting systems
- Information gathering systems

Some particular IWSN application areas include area monitoring, machine health monitoring, structural monitoring, air/water quality, and waste monitoring [5]. With significant automation, applications tend to reinstate legacy monitoring systems with more intelligent applications that can react to stimuli received in the form of sensed data. Moreover, these applications are expected to learn and auto-tune their behavior as circumstances vary from time to time. Sensor and actor networks demonstrate this behavior in the form of emergency response, disaster prevention, industrial automation, feedback control, etc.

1.1.2 WSNs and IoT

Historically, WSNs have been domain-specific with tailor-made solutions, deployed to serve a particular application scenario. Recently, as the capabilities (e.g., sensing, processing, communication, energy replenishment, etc.) of the sensor nodes increase at scale, these are becoming more and more connected, integrating to realize the IoT. Some common examples include smart homes, buildings, and cities that promote the integration of diverse WSN deployments. In such scenarios, applications like appliance monitoring, surveillance, water management, heating, ventilation, air conditioning, and elevator control are integrated to form the IoT [6].

1.2 ADAPTIVITY AND QoS CHALLENGE

The sensor nodes are battery-powered and often deployed at remote places, making them difficult to replace or replenish. This requires purposive energy consumption and exploration of up-to-the-minute WSN design [7]. However, the incremental diversification in deployment domains and the variation of applications put stringent and assorted quality-of-service (QoS) requirements on WSNs [5]. In addition to energy, factors like reliability, delay, throughput, and signal strength have become of utmost importance over time [8, 9].

1.2.1 LEGACY SOLUTIONS

To tackle the QoS challenge, most research has focused on devising new medium access control (MAC) [9] or routing protocols [10], and cross-layer approaches [11]. An intractable optimization problem, also regarded as NP-hard, needs to be devised to solve an optimization QoS problem at hand. Moreover, confronting applications where multiple QoS metrics have to be considered simultaneously, facilitating QoS becomes a multi-objective optimization (MOO) problem, adding adversely to the computational complexity [12]. The solutions for MOO problems are often based on approximations, trading-off accuracy for the sake of complexity. Further, the legacy approaches lack in yielding a clear solution that can assist adaptation and conformity when the network evolves and changes occur.

1.2.2 DATA-DRIVEN SOLUTIONS

To devise a solution accommodating multiple QoS metrics with sufficient accuracy and inculcating adaptivity as the network changes, data-driven approaches have become popular. Such systems use performance data coupled with intelligent algorithms based on artificial intelligence (AI) and machine learning (ML) to accurately predict the QoS metrics [13]. Intelligence, adaptivity, affordable complexity, and availability of performance data are among the prominent factors driving the adoption of data-driven techniques for QoS predictions in WSNs.

1.3 THE DATA-DRIVEN PARADIGM

Two prominent features of modern computing systems are intelligence and adaptivity, driven by AI and ML. The communication domain is no different in this regard. Data-driven networking (DDN) has been proposed with centralized control [14]. Using the data-driven approach, WSNs can learn their performance recordings and then harmonize and fine-tune their own responses in accordance with the QoS requirements.

The sole purpose of using the centralized approach is to enable sophisticated algorithms to run on a substantial volume of data without putting the load on the constrained sensor nodes. A typical flow of the potential data-driven paradigm for WSNs is shown in Figure 1.1. The performance data is collected from heterogeneous

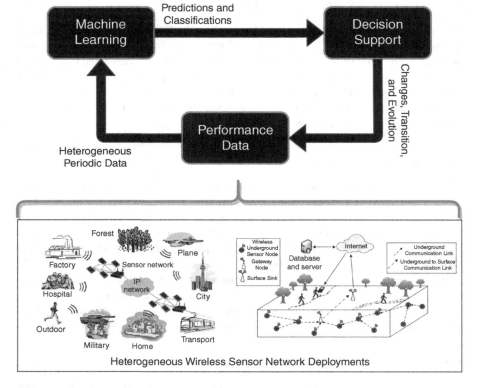

FIGURE 1.1 The data-driven paradigm.

deployments of WSNs and centrally processed using techniques like ML to learn the QoS attributes based on various characteristics and parameters of the network. The prediction results are used for making intelligent decisions to meet desired QoS goals. This learning loop continues as the transitions in different settings of the WSNs occur and the system evolves.

1.3.1 QoS Metrics and Adaptivity

In wireless, QoS is considered in terms of timeliness in which bandwidth requirements and reliability are key. In addition to these, no QoS system can be realized without earnest consideration for energy in the case of WSNs [15]. Moreover, adaptivity is a characteristic of increasing importance in modern communication systems. QoS metrics and their relationships combined with an adaptivity loop are shown in Figure 1.2. Next, these QoS factors are briefly explained.

1.3.1.1 Timeliness

Timeliness is measured primarily using the delay as a metric. Depending on the particular context under consideration, the definitions of delay can vary slightly.

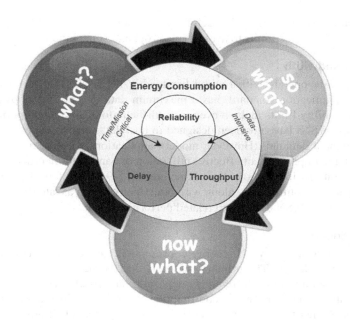

FIGURE 1.2 QoS metrics, their relationships, and adaptivity.

In the norm of this thesis, the delay is defined as the time between the first transmission from sender and reception at receiver including retransmission, if any. Delay is an important metric for many application domains where certain time constraints apply in decision-making [16]. Breach of time-limit can result in different kinds of losses including safety and finance. Such cases occur in health care, IWSNs, and any emergency response situations.

Other metrics and terms frequently used to measure timeliness include latency, response time, jitter, etc.

1.3.1.2 Reliability

Reliability is of paramount importance in the performance of any wireless communication system because of its inherent vulnerability to interference and error-prone links. Moreover, standards like IEEE 802.15.4 use the industrial, scientific, and medical (ISM) band for low-power communication. Since other standards, such as WiFi use the same bands with a significantly higher power, achieving reliability can be challenging and power consumption becomes a major concern as well. At the physical (PHY) layer, bit error rate (BER) is used as an indicator of reliability. However, at higher layers, the packet delivery ratio is used since data is grouped in the form of packets.

The packet delivery ratio is important for effective monitoring and response. The decrease in packet delivery ratio does not just downgrade the performance of WSN, but also causes a hike in energy consumption due to retransmissions. Coupled with delay, the packet delivery ratio defines an important class of applications, regarded as time- and mission-critical. In such settings, data must be reliably delivered by observing certain time limits to avoid a catastrophe of some kind [17].

1.3.1.3 Bandwidth

WSNs typically operate with limited bandwidth. IEEE 802.15.4, a popular low-power communication standard, has a maximum data rate of 250 Kbps [18]. The inherent reliability limits make it even more difficult to achieve good performance. Historically, WSNs have not been designed to host data-intensive applications with high throughput demands. However, modern applications with multimedia content bring forth the importance of effective utilization of bandwidth in order to facilitate critical operations like remote patient monitoring [19]. Therefore, bandwidth and usage efficiency requirements, along with other metrics such as latency and reliability, are critical to WSN design and related performance goals [20].

1.3.1.4 Energy

Since the emergence of WSN, energy consumption has been a predominant consideration in its design. The primary reason is that the sensor nodes are generally battery-powered and often so remotely deployed that repeated access to the nodes is not possible. Therefore, energy consumption has to be considered alongside other QoS goals. Some of the metrics associated with energy include lifetime maximization, transmission power control, and energy consumption minimization [7, 21]. In this book, we have used energy consumed per useful bit delivered as the metric to predict.

1.3.1.5 Adaptivity

Adaptivity is a trait that all modern computer systems should enact as they undergo changes and evolution. Although the term "adaptive" has frequently been used for different research dimensions in the domain of WSNs, the focus on the real definition of an adaptive QoS system has lacked a great deal [11, 22, 23]. To be truly adaptive, a communications system must be able to inculcate changes automatically to facilitate QoS where inputs and other diverse characteristics of the systems change/evolve. Such a behavior is also regarded as self-organizing and self-optimizing. Thus, adaptivity operates in a loop that continues to collect and observe real-time data and intelligently learns influential and conclusive relationships and patterns, hence, providing decision optimizing support to meet desired QoS objectives [14].

REFERENCES

1. Akyildiz, Ian F., Weilian Su, Yogesh Sankarasubramaniam, and Erdal Cayirci. "Wireless sensor networks: A survey." *Computer networks* 38, no. 4 (2002): 393–422.
2. Rawat, Priyanka, Kamal Deep Singh, Hakima Chaouchi, and Jean Marie Bonnin. "Wireless sensor networks: A survey on recent developments and potential synergies." *The journal of supercomputing* 68, no. 1 (2014): 1–48.
3. Arampatzis, Th, John Lygeros, and Stamatis Manesis. "A survey of applications of wireless sensors and wireless sensor networks." In *Proceedings of the 2005 IEEE International Symposium on Intelligent Control, Mediterranean Conference on Control and Automation*, pp. 719–724. IEEE, 2005.

4. Zand, Pouria, Supriyo Chatterjea, Kallol Das, and Paul Havinga. "Wireless industrial monitoring and control networks: The journey so far and the road ahead." *Journal of sensor and actuator networks* 1, no. 2 (2012): 123–152.

5. Raza, Mohsin, Nauman Aslam, Hoa Le-Minh, Sajjad Hussain, Yue Cao, and Noor Muhammad Khan. "A critical analysis of research potential, challenges, and future directives in industrial wireless sensor networks." *IEEE communications surveys & tutorials* 20, no. 1 (2017): 39–95.

6. Gubbi, Jayavardhana, Rajkumar Buyya, Slaven Marusic, and Marimuthu Palaniswami. "Internet of Things (IoT): A vision, architectural elements, and future directions." *Future generation computer systems* 29, no. 7 (2013): 1645–1660.

7. Rault, Tifenn, Abdelmadjid Bouabdallah, and Yacine Challal. "Energy efficiency in wireless sensor networks: A top-down survey." *Computer networks* 67 (2014): 104–122.

8. Chen, Dazhi, and Pramod K. Varshney. "QoS support in wireless sensor networks: A survey." In *International conference on wireless networks*, vol. 233, pp. 1–7. 2004.

9. Yigitel, M. Aykut, Ozlem Durmaz Incel, and Cem Ersoy. "QoS-aware MAC protocols for wireless sensor networks: A survey." *Computer networks* 55, no. 8 (2011): 1982–2004.

10. Yetgin, Halil, Kent Tsz Kan Cheung, Mohammed El-Hajjar, and Lajos Hanzo. "A survey of network lifetime maximization techniques in wireless sensor networks." *IEEE communications surveys & tutorials* 19, no. 2 (2017): 828–854.

11. Al-Anbagi, Irfan, Melike Erol-Kantarci, and Hussein T. Mouftah. "A survey on cross-layer quality-of-service approaches in WSNs for delay and reliability-aware applications." *IEEE Communications surveys & tutorials* 18, no. 1 (2014): 525–552.

12. Fei, Zesong, Bin Li, Shaoshi Yang, Chengwen Xing, Hongbin Chen, and Lajos Hanzo. "A survey of multi-objective optimization in wireless sensor networks: Metrics, algorithms, and open problems." *IEEE communications surveys & tutorials* 19, no. 1 (2016): 550–586.

13. Jiang, Junchen. *"Enabling Data-Driven Optimization of Quality of Experience in Internet Applications."* PhD diss., Intel Corporation, 2017.

14. Jiang, Junchen, Vyas Sekar, Ion Stoica, and Hui Zhang. "Unleashing the potential of data-driven networking." In *International Conference on Communication Systems and Networks*, pp. 110–126. Springer, Cham, 2017.

15. Chen, Ray, Anh Phan Speer, and Mohamed Eltoweissy. "Adaptive fault-tolerant QoS control algorithms for maximizing system lifetime of query-based wireless sensor networks." *IEEE Transactions on dependable and secure computing* 8, no. 2 (2010): 161–176.

16. Kim, Joohwan, Xiaojun Lin, Ness B. Shroff, and Prasun Sinha. "Minimizing delay and maximizing lifetime for wireless sensor networks with anycast." *IEEE/ACM transactions on networking* 18, no. 2 (2009): 515–528.

17. Mahmood, Muhammad Adeel, Winston KG Seah, and Ian Welch. "Reliability in wireless sensor networks: A survey and challenges ahead." *Computer networks* 79 (2015): 166–187.

18. Lu, Gang, Bhaskar Krishnamachari, and Cauligi S. Raghavendra. "Performance evaluation of the IEEE 802.15. 4 MAC for low-rate low-power wireless networks." In *IEEE International Conference on Performance, Computing, and Communications*, 2004, pp. 701–706. IEEE, 2004.

19. Tung, Hoi Yan, Kim Fung Tsang, Hoi Ching Tung, Kwok Tai Chui, and Hao Ran Chi. "The design of dual radio ZigBee homecare gateway for remote patient monitoring." *IEEE Transactions on consumer electronics* 59, no. 4 (2013): 756–764.

20. Jabbar, Sohail, Abid Ali Minhas, Muhammad Imran, Shehzad Khalid, and Kashif Saleem. "Energy efficient strategy for throughput improvement in wireless sensor networks." *Sensors* 15, no. 2 (2015): 2473–2495.

21. Ren, Ju, Yaoxue Zhang, Kuan Zhang, Anfeng Liu, Jianer Chen, and Xuemin Sherman Shen. "Lifetime and energy hole evolution analysis in data-gathering wireless sensor networks." *IEEE transactions on industrial informatics* 12, no. 2 (2015): 788–800.

22. Ezdiani, Syarifah, Indrajit S. Acharyya, Sivaramakrishnan Sivakumar, and Adnan Al-Anbuky. "An IoT environment for WSN adaptive QoS." In *2015 IEEE International Conference on Data Science and Data Intensive Systems*, pp. 586–593. IEEE, 2015.

23. Ezdiani, Syarifah, Indrajit S. Acharyya, Sivaramakrishnan Sivakumar, and Adnan Al-Anbuky. "Wireless sensor network softwarization: Towards WSN adaptive QoS." *IEEE internet of things journal* 4, no. 5 (2017): 1517–1527.

2 A Collaborative Data-Driven Intelligence for Future Wireless Networks

Rashid Ali
Department of Software Device Engineering,
Sejong University, Seoul, Republic of Korea

Hyung Seok Kim
College of Software Convergence, Sejong
University, Seoul, Republic of Korea

CONTENTS

2.1 INTRODUCTION

Fifth generation (5G) and beyond 5G (B5G) networks are promising to cooperate with future wireless local area networks (WLANs) like their incumbent technologies that provided inspiring services, such as astonishingly high throughput services. The amalgamation of WLAN (Wi-Fi) networks with 5G/B5G networks has been a topic of interest for researchers over the past two decades. According to a

DOI: 10.1201/9781003216971-3

recent report [1], more than 70% of mobile data traffic is generated by WLAN networks. Next-generation ultra-dense WLANs invite significant commitment from the research community to enable such promising services. IEEE working group (WG) has recently launched an amendment to IEEE 802.11 WLANs, named IEEE 802.11ax high-efficiency WLAN (HEW) [2]. HEW deals with ultra-dense and various device deployment scenarios for 5G/B5G, such as sports stadiums, train stations, and shopping centers. It is anticipated that HEW infers the exciting features of both the device's environment as well as the device's interacting behavior with its environment to spontaneously manage the medium access control (MAC) layer resource allocation parameters.

2.2 DATA-DRIVEN WIRELESS NETWORKS AND INTELLIGENCE

Recently, data has played a vital role for the design and development of network architecture and solutions. Wireless networks continuously generate data for the purpose of spectral efficiency. Much of the generated data are discarded without further use, such as the RSSI values, interference levels, and backoff counters. Such a data generated within the wireless networks for efficient spectrum allocation may be used to perform intelligent mechanisms for the optimization of the process. Practically, a WLAN device, also referred to as a station (STA), proficiently and dynamically manages the wireless channel resources, such as the MAC layer's distributed coordination function (DCF), which utilizes a carrier sense multiple access with collision avoidance (CSMA/CA) mechanism to improve the STA's quality of experience (QoE). In general, STA performance relies upon the exploitation of the uncertainty of the network heterogeneity in terms of traffic variety. Traditional IEEE 802.11 WLANs use a binary exponential backoff (BEB) method as a CSMA/CA mechanism to maintain a near collision-free environment. BEB uses a random backoff value picked from a contention window (CW) to contend for the wireless medium resources. The initial and maximum sizes of CW are fixed by the standardization, and an STA exponentially increases the initial size of CW after every time it encounters a collision (collision in the WLAN is assumed if acknowledgment of any transmission is not received) until it reaches the maximum size of the CW. The size of a CW is always set back to its initial value once a packet is transmitted successfully. However, this blind increase/decrease of contention parameters leads to performance degradation in the network, such as for a highly dense network situation; resetting the size of the CW to its initial size (minimum size) may induce more collisions due to a smaller CW size for a more substantial number of contenders. Similarly, if the network density is minimal, for example, only a few STAs are contending for the medium resources, the exponential increase of CW size for coincidental collisions causes unnecessary delays in the network. WLAN resources are fundamentally constrained due to shared medium access infrastructure, while its services have become progressively sophisticated and diverse. Therefore, to accomplish the targeted objectives of HEW, it is imperative to investigate efficient and robust resource allocation schemes.

Recently, reinforcement learning (RL) has prospered to empower machine intelligence (MI) capabilities in wireless networks. RL is currently a flourishing technique in active research areas into relevant use cases of 5G and B5G systems, ranging from learning complex situations with unfamiliar channel models to the deployment of 5G new radio (5G NR) networks [3]. It is inspired by the behaviorist learning and control theory, where a device can accomplish a goal by collaborating with its environment. RL utilizes explicit learning algorithms, such as Q-learning (QL) to solve the Markov decision process (MDP) models [4]. It uses these models in applications like learning an unfamiliar wireless network environment and resource allocations in ultra-dense WLANs [5].

2.3   CONTRIBUTIONS OF THIS CHAPTER

Persuaded by the potential applications and features of RL techniques in wireless networks, the authors in [5] present RL as a framework for MAC layer resource allocation in dense WLANs. Their proposed RL framework utilizes QL-based inference to optimize the performance of the contending STAs, and they called it an intelligent QL-based resource allocation (iQRA) mechanism. Authors show that iQRA optimizes the performance of the BEB method by utilizing an observation-based channel collision probability (p_{obs}) [6]. In the iQRA mechanism, each STA manages the contention parameters based on the p_{obs}, which is iteratively optimized with the help of the QL algorithm. However, in their proposed mechanism, every STA must optimize its contention parameters based on its observed and accumulated Q-value from the value function (we will briefly discuss Q-value and value function in later sections). A wireless network environment is of distributed and dynamic nature, which changes more often. Thus, relying on the individual estimations and optimizations may lead to higher error variance. Therefore, in this chapter, we propose a federated RL (FRL) framework to collectively and cooperatively optimize the network contention parameters in ultra-dense WLANs. In addition, we propose four potential applications of FRL framework for incumbent technologies in 5G/B5G networks.

2.4   COLLABORATIVE REINFORCEMENT LEARNING FRAMEWORK

Federated learning (FL), also known as collaborative learning, is an ML technique that learns the environment across numerous decentralized devices without sharing their data. This ML technique is very different from the traditional centralized ML techniques, where all the target data is needed to be uploaded to a single server. Moreover, FL empowers numerous agents to construct a typical, strong ML model without sharing the actual data/information. As a decentralized ML technique, FL tends to the security and privacy concerns by disseminating the information to distributed agents in the environment. FL uses distributed learning, which is a compelling technique to acknowledge collaborative learning. The applications of FL are already recognized by several research areas, such as 5G/B5G, IoT, and blockchain networks [7–12]. Therefore, in this chapter, we extend the capabilities of FL techniques to utilize in RL models, named the FRL model.

   In this section, we articulate our proposed FRL model for resource allocation mechanisms in HEW networks. This section is further divided into three subsections. The first subsection overviews the RL with its important elements. In the second subsection, we discuss the existing iQRA mechanism for resource allocations in dense WLANs. The third subsection elaborates on our proposed FRL model to further optimize the performance of the iQRA mechanism.

2.5   REINFORCEMENT LEARNING AS A FRAMEWORK

In RL, an STA iteratively learns the behavior of its actions (a_t) at time t and maps them to the prospective decisions to maximize its reward (r_t). Here, the reward is a numerical valued response of the action from the environment at a specific state (s_t). Typically, a learning STA in RL does not have the foggiest idea what actions to perform, yet it must discover optimal actions (a_t^*) to achieve the best reward from the environment. Here, the primary goal is to maximize the accumulated reward collected in the long run. Therefore, a reward of the action articulates how pleased an STA is in any specific state. Hence, the reward is the key motivation for changing the policy (π) at any state, that is, the strategy of selecting low-reward actions might be changed to choose other actions for a specific state in the future. Another important element of the RL technique is a value function, also referred as a Q-value function $Q(s_t, a_t)$. A reward of the action represents what is good in an immediate sense. However, a value function shows the accumulated reward as what is best in the end [5].

Consequently, the q-value function of a state is the accumulated reward of an STA collected in the long run, starting from the very first state. For example, a state may consistently yield a low reward yet, simultaneously, have a high Q-value function if it is regularly trailed by different states that produce high rewards. Regardless, it is the Q-value with which we are most concerned when making and evaluating decisions. Thus, we always seek the actions that yield Q-value, not the most reward, since these actions bring the highest extent of rewards over the long run.

One of the challenges faced by the RL is the trade-off between exploration and exploitation. To get a significant reward, an RL STA must learn toward actions endeavored previously and check to be compelling in making a reward. However, to discover such actions, it needs to attempt actions that it has not chosen previously; that is what is known as exploration. Likewise, the STA requires to exploit what it has efficiently explored, keeping in mind that the target objective is to obtain the maximized accumulated reward (known as exploitation). Figure 2.1 portrays an example of RL with its core elements in a WLAN environment.

2.5.1 iQRA MECHANISM

The RL-based iQRA mechanism considers backoff stages as an available finite set of m potential states, $S = \{0, 1, 2, \ldots, m\}$ where a learning STA increases the size of CW by moving to the next state and decreases the size of CW by moving back to the previous state. An action a_t, in a specific state s_t, receives a reward r_t at time t, with the objective to exploit its accumulated Q-value function, $Q(s, a)$.

One of the main objectives of the iQRA mechanism is to minimize WLAN channel collision probability, which is p_{obs}. Therefore, the reward (r_t) given by a_t taken at s_t is formulated as, $r_t(s_t, a_t) = 1 - p_{obs}$. An STA observes its current state (s_t), that is $s_t = s \in S$, and takes an action (a_t), that is $a_t = a \in A$. This action moves the STA to

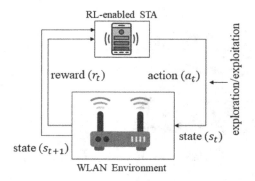

FIGURE 2.1 An example of RL with its core elements in a WLAN environment.

$s_{t+1} = s' \in S$. The *i*QRA mechanism aims to find an optimal policy that exploits the accumulated Q-value ($Q_t(s_t, a_t)$), which is updated as follows:

$$Q_t(s_{t+1}, a_{t+1}) = (1 - \alpha) \times Q_t(s_t, a_t) + \alpha \times \Delta Q_t(s_t, a_t) \tag{2.1}$$

This Q-value function is updated iteratively after the STA performs an action and perceives the feedback reward. Here, $0 < \alpha < 1$ is a learning-rate control parameter. In Equation 2.1, learning estimate ΔQ is updated as follows here:

$$\Delta Q_t(s_t, a_t) = \left\{ r_t(s_t, a_t) + \beta \times \max_{a^*} Q_t(s', a') \right\} - Q_t(s_t, a_t) \tag{2.2}$$

Where parameter β is known as a discount factor and is expressed as $0 < \beta < 1$ and weighs instant reward more aggressively than future reward. The $\max_a Q_t(s', a')$ represents the best estimated Q-value for the future state-action pair.

The RL-based *i*QRA mechanism utilizes this optimized Q-value to determine optimal contention parameters, which is a selection of the current backoff stage and the size of the current CW. In the *i*QRA mechanism, optimization of contention parameters is performed individually at each STA, where high error variance may occur and, thus, can lead to more severe estimations. Therefore, we propose to collaborate the accumulated Q-value in the network along with other STAs to reduce estimation error variance.

2.6 PROPOSED DATA-DRIVEN FRL FRAMEWORK

The FRL system model is outlined in Figure 2.2. We assume a WLAN access point (AP) as a centralized device of the WLAN environment. STAs around the AP have their individual learning models (such as RL-based *i*QRA) to optimize the channel access parameters for the transmission to and from the AP (that is estimated Q-value, $Q_t(s_t, a_t)$). For this purpose, STAs sense the channel for observation-based collision probability p_{obs} as formulated by the *i*QRA. In the *i*QRA mechanism, the competing STAs perform a BEB procedure for channel resources with the selection of a random backoff value B after the channel is sensed idle for a DIFS period, as

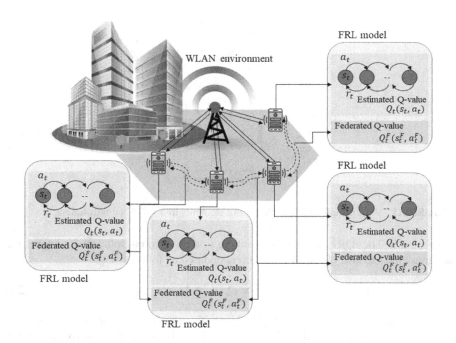

FIGURE 2.2 Proposed FRL frameworks for resource allocation in 5G/B5G WLANs.

shown in Figure 2.3 ($B = 9$ for STA 1, and $B = 7$ for STA 2). The discretized time slots during the BEB procedure are observed as either idle ($S_k = 0$) or busy ($S_k = 1$), where $k = \{0, 1, 2, 3, \ldots, B_{obs} - 1\}$ for B_{obs} total number of observed backoff time slots. As shown in Figure 2.4, an STA in a WLAN environment formulates p_{obs} as follows, $\frac{1}{B_{obs}} \times \sum_{k=0}^{B_{obs}-1} S_k$ [5]. The formulated p_{obs} works as a reward for the STA to accumulate its Q-value function further, which is Equation (2.1) and Equation (2.2), respectively. Since individual estimated Q-value may suffer from large overestimation (due to error variances), in the proposed FRL model, every STA integrates its accumulated Q-value ($Q_t(s_t, a_t)$) in the acknowledgment (ACK) packet, known as federated ACK (FACK) message to send it to other active STAs in the WLAN. This federated Q-value ($Q_t(s^F, a^F)$) provides a second Q function to the QL algorithm (as shown in Figure 2.3), which is also known as double QL (DQL) method [13]. In Equation (2.2), action a' shows the maximum valued action for prospective state s', according to the Q-value function $Q_t(s_t, a_t)$. However, instead of using $\max_{a^*} Q_t(s', a')$ to update Q-value, as QL algorithm would do, FRL uses the value $Q_t(s^F, a^F)$. Since federated Q-value Q^F was updated on the same WLAN environment, but with a different set of observations, it represents a fair estimate for the Q-value of this action. For the FRL model, both Q-value functions (estimated and federated) must learn from separate sets of experiences in the same environment. However, an STA uses both Q-value functions to update its local Q-value estimate as follows:

$$\Delta Q_t(s_t, a_t) = \left\{ r_t(s_t, a_t) + \beta \times Q_t\left(s^F, a^F\right) \right\} - Q_t(s_t, a_t) \qquad (2.3)$$

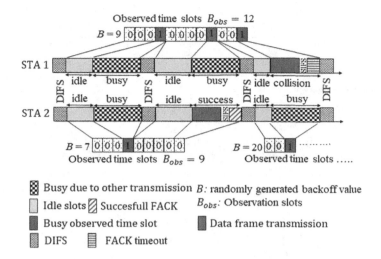

FIGURE 2.3 FRL-based ACK (FACK) collaboration methods during the CSMA/CA Mechanism.

2.7 CONVERGENCE EVALUATION

We evaluated the proposed scheme with much iteration for the convergence evaluation. Figure 2.4 shows a Q-value estimate (ΔQ) convergence comparison between a non-federated RL (QL) algorithm and a federated RL (DQL) algorithm for 1,000 iterations. We observe that the FRL converges faster than RL, which is evident in the performance enhancement. A rapid convergence helps STAs to learn their WLAN environment swiftly and allows them to optimize their resource allocations. Instinctively, this is what we would expect: the RL mechanism depends on the individual estimator (Q-value accumulator), and the FRL mechanism depends on the double estimator (individual and federated).

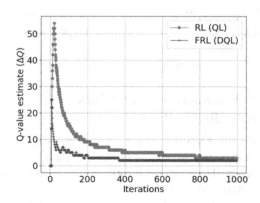

FIGURE 2.4 Q-value estimate (ΔQ) convergence of FRL (DQL) and RL (QL) algorithms.

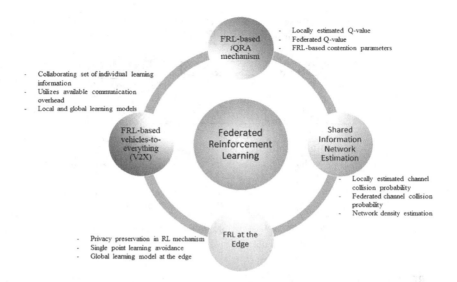

FIGURE 2.5   Potential applications of FRL-based framework.

2.8   POTENTIAL APPLICATIONS

A few of the potential applications of our proposed FRL model are presented in this section. In the future, we aim to implement our proposed model for these applications. Figure 2.5 articulates our proposed potential applications of the FRL technique.

2.9   FRL-BASED *i*QRA (F*i*QRA) MECHANISMS

In the *i*QRA mechanism, STAs optimize their contention parameters based on their individually estimated Q-value. The QL algorithm performs unwell in numerous dynamic and stochastic WLAN environments. This performance degradation is mainly caused by substantial overestimation of Q-values due to the use of $\max_{a} Q_t(s', a')$ in the QL algorithm. Therefore, to cure this issue, we propose to use the FRL-based *i*QRA (F*i*QRA) mechanism. In the F*i*QRA mechanism, an STA competing for the channel resources is expected to learn the WLAN environment faster than an STA in *i*QRA mechanism, as evaluated in Figure 2.4.

2.10   SHARED INFORMATION NETWORK
ESTIMATION (SINE) METHOD

The performance of a WLAN environment strongly relies upon the number of contending STAs, n, which is the number of devices that are simultaneously trying to access channel resources. However, the information of n competing for the channel resources cannot be retrieved even in the presence of an AP due to a limited number of associated STAs (which are usually different than the actual active STAs in the WLAN environment). The ability to estimate n induces numerous implications for an STA in a WLAN environment. In [14], authors indicate that the BEB CW depends upon n to maximize the WLAN system performance. However, in a distributed WLN environment, the accuracy of the individually estimated n degrades as the number of STAs increases due to the amplified errors in the collision probability (error variance) [14]. Also, due to the non-linear relation between n and channel collision probability, the individual estimation of n is unfair. Therefore, we also propose to utilize the FRL-based shared information network estimation (SINE) method. The SINE method uses a FACK message to collaborate an estimated n value. Further, the CW of an STA can be resized based on the estimated n.

2.11 FRL AT THE EDGE (FEdge)

Recently, edge computing (EC) has seen increased enthusiasm due to its capacity to broaden cloud computing efficiencies to the wireless network edge with low latency. Various 5G/B5G applications, such as autonomous driving cars, augmented reality (AR), remote surveillance, and tactile Internet with low latency. In such applications, the connected edge user devices have rigorous computational resource constraints. One approach to furnish those edge devices with on-demand computing resources is to utilize a cloud network. However, the characteristic deferral relating to end-to-end communications with a cloud server can prompt to intolerable latency. Therefore, it is essential to utilize FRL at the edge (FEdge) to empower different insightful applications. Conventional RL uses centralized or individual learning data, which requires transferring of information from numerous geographically distributed devices to the learning device. It is important to introduce collaborated and edge-deployed RL techniques, such as FRL, to cope with the challenge of environment data privacy preservation in RL mechanisms. FRL enables data privacy preservation in the network by avoiding the use of centralized or single-point learning. FEdge involves a set of wireless devices within its environment and computation of the global learning model at the edge of the network.

2.12 FRL-BASED VEHICLE-TO-EVERYTHING (V2X) COMMUNICATIONS

In 5G and B5G networks, vehicle-to-everything (V2X) communication is a key enabler for self-driving autonomous cars. V2X communication assists in expanding the effectiveness of transportation frameworks by employing collaborative task handling. However, the performance of autonomous services in a self-driving car, such as instant navigation, collision evasion, and collaborative task handling, significantly depends on the capacity to communicate with ultra-reliable and low latency communications (URLLC). The target end-to-end (E2E) latency for URLLC requirement is of < 1 ms. To achieve this targeted latency requirement, most of the existing related work focuses on improving the expected latency of a V2X network by utilizing a probabilistic control to sustain queuing delays at the end devices. However, a probabilistic control approach may improve network reliability on shorter queue length devices and it fails to control the exceptional actions of large queue-length devices with low channel access probability. Therefore, few of the end devices (vehicles) in the V2X network may encounter volatile latencies causing network performance degradation. The major issue with the use of probabilistic methods is due to the lack of enough information samples of rare extraordinary events.

In V2X communication, roadside units (RS Us) can help vehicles gather numerous information samples over the network at the cost of additional communication overheads. As we mentioned earlier, FRL permits learning models by collaborating a set of individual/local learning information with other devices within available communication overheads (such as the use of FACK messages). Besides, the FRL framework does not depend on synchronization among other devices in a V2X network due to locally implemented QL algorithm. Thus, even if connectivity between the vehicle and RSU fails, a vehicle can, in any case, learn and explore the environment with the help of locally available information.

2.13 CONCLUSIONS

The behaviorist learning technique of RL has helped to enable machine intelligence in wireless communication networks, such as 5G and B5G. RL is currently an emerging ML technique in numerous active research areas of 5G and B5G. It employs explicit ML algorithms, such as Q-learning to solve MDP models in applications like learning an unfamiliar wireless network environment, and resource allocations in ultra-dense WLANs. A wireless network environment is dynamic, meaning it continuously changes. RL algorithms relying on the individual estimations and optimizations may lead to higher error variance. Therefore, in this chapter, we propose a FRL model for ML-enabled resource allocations in ultra-dense 5G wireless networks, such as IEEE 802.11ax WLANs, to overcome the wireless channel collision issues. The experiment results of non-federated RL and proposed FRL signify the prominence of our proposed model. We also highlight four of the potential applications of the proposed FRL model for incumbent technologies in 5G/B5G: FRL-based iQRA (FiQRA) mechanism, SINE method, FEdge, and FRL-based V2X communication. In future work, we aim to implement our proposed FRL model for these applications and to contribute to the research community. We encourage researchers from institutions as well as from industry to consider our proposed FRL model for their potential research.

REFERENCES

1. M. Dano, "How Much Cellular and Wi-Fi Data Are Smartphone Users Consuming, and With Which Apps?" FierceWireless, Jan. 24, 2017. [Online]. Available: https://www. fiercewireless.com/wireless/how-much-cellular-and-wi-fi-data-are-smartphone-users-consuming-and-which-apps-verizon-0 Accessed on: Jan. 20, 2020.
2. IEEE802.org, "Status of Project IEEE 802.11ax," 2018. [Online]. Available: http://www.ieee802.org/11/Reports/tgax_update.htm Accessed on: Jan. 1, 2020.
3. 3GPP.org. "Study on New Radio (NR) access technology," Specification 38.912. Release 14. 2017. [Online]. Available: https://portal.3gpp.org/desktopmodules/Specifications/SpecificationDetails.aspx?specificationId=3059
4. R. S. Sutton and A. G. Barto, Reinforcement Learning: An Introduction, 2nd ed. Cambridge, MA, USA: MIT Press, 1998.

5. R. Ali, N. Shahin, Y. B. Zikria, B. Kim, and S. W. Kim, "Deep Reinforcement Learning Paradigm for Performance Optimization of Channel Observation-Based MAC Protocols in Dense WLANs," IEEE Access, vol. 7, pp. 3500–3511, 2019. DOI: 10.1109/ACCESS.2018.2886216

6. R. Ali, N. Shahin, Y. Kim, B. Kim, and S. W. Kim, "Channel Observation-Based Scaled Backoff Mechanism for High-Efficiency WLANs," IET Electronics Letters, vol. 54, no. 10, pp. 663–665, May 2018.

7. Z. Han, T. Lei, Z. Lu, X. Wen, W. Zheng, and L. Guo, "Artificial Intelligence-Based Handoff Management for Dense WLANs: A Deep Reinforcement Learning Approach," in IEEE Access, vol. 7, pp. 31688–31701, 2019. DOI: 10.1109/ACCESS.2019.2900445

8. D. Loghin et al., "The Disruptions of 5G on Data-driven Technologies and Applications," in IEEE Transactions on Knowledge and Data Engineering, vol. 32, no. 6, pp. 1179–1198, 2020. DOI: 10.1109/TKDE.2020.2967670

9. S. Savazzi, M. Nicoli and V. Rampa, "Federated Learning with Cooperating Devices: A Consensus Approach for Massive IoT Networks," in IEEE Internet of Things Journal, vol. 7, no. 5, pp. 4641–4654, 2020. DOI: 10.1109/JIOT.2020.2964162

10. Y. Lu, X. Huang, Y. Dai, S. Maharjan, and Y. Zhang, "Differentially Private Asynchronous Federated Learning for Mobile Edge Computing in Urban Informatics," in IEEE Transactions on Industrial Informatics, vol. 16, no. 3, pp. 2134–2143, March 2020. DOI: 10.1109/TII.2019.2942179

11. J. Kang, Z. Xiong, D. Niyato, H. Yu, Y. Liang and D. I. Kim, "Incentive Design for Efficient Federated Learning in Mobile Networks: A Contract Theory Approach," 2019 IEEE VTS Asia Pacific Wireless Communications Symposium (APWCS), Singapore, 2019, pp. 1–5. DOI: 10.1109/VTS-APWCS.2019.8851649

12. M. Yao, M. Sohul, V. Marojevic, and J. H. Reed, "Artificial Intelligence Defined 5G Radio Access Networks," in IEEE Communications Magazine, vol. 57, no. 3, pp. 14–20, March 2019. DOI: 10.1109/MCOM.2019.1800629

13. H. van Hasselt, A. Guez, and D. Silver, "Deep Reinforcement Learning with Double Q-learning," in Proc. Of the 30th AAAI Conference on Artificial Intelligence, Phoenix, AZ, 2016, pp. 2094–2100.

14. G. Bianchi and I. Tinnirello, "Kalman Filter Estimation of the Number of Competing Terminals in an IEEE 802.11 Network," IEEE INFOCOM 2003. Twenty-second Annual Joint Conference of the IEEE Computer and Communications Societies, San Francisco, CA, 2003, pp. 844–852 vol.2. DOI: 10.1109/INFCOM.2003.1208922

3 Federated Learning Technique in Enabling Data-Driven Design for Wireless Communication

Ahmad Arsalan
Department of Computer Science and Information
Technology, University of Central Punjab, Pakistan

Tariq Umer
Department of Computer Science, COMSATS University,
Lahore Campus, Pakistan

Rana Asif Rehman
Department of Computer Science, National University of
Computer and Emerging Sciences, Lahore Campus, Pakistan

CONTENTS

DOI: 10.1201/9781003216971-4

3.1   INTRODUCTION

Data-driven research, which is more commonly known as "data science," is a type of study that focuses on extracting information from huge data sets to address a specific problem. Data science has grown in prominence to gain a better understanding of complex systems' behavior that are difficult to predict and model. As a result, an article offers a potential meaning of data science, which is defined as "the study of generalizable information extraction from data" [1]. A more straightforward definition may be that data science facilitates the discovery and extraction of novel insights from large amounts of data. With this, data-driven techniques deal with vast volumes of experimental data in data sets, and those techniques are used to build models that may be utilized to better analyze system behavior, extract new knowledge, and generate predicted data that closely resembles the actual observed data. The analysis usually involves an excessive amount of data, resulting in many gigabytes of traces. Data-driven research is becoming more popular in a variety of sectors [2]. The study of the human genome to predict an individual's susceptibility to certain diseases [3], evaluation of social media connections [4], client purchase history [5], applications of cloud computing [6, 7], and mobile cellular networks [8] are all possible scenarios where data science applications are being used. A wireless network is an attractive data science application due to its unpredictability. As a result, both natural and man-made events have an impact on it. On one hand, it relies on electromagnetic transmission, and on the other, it also uses network technology, which comprises human-made hardware and software components. Therefore, the implementation of a wireless network can be done in simulations, but it is difficult to recreate in the real world. Because of these constraints, simulations alone cannot be used to identify and/or explain diverse wireless system characteristics. For example, both network and physical layers have an impact on the interval between arrivals of wireless data packets. In wireless networks, data science is used to model complex systems, network utilization, and predict system behavior. The recently suggested federated learning (FL) framework [9] has gotten a lot of interest since it allows many mobile users to work together to train machine learning (ML) models on a wide scale. FL is a new decentralized system technique that considers different issues such as privacy and security, data distribution, and resources utilization. It uses a device's processing power and previous data to decentralize model formation and store data where it is generated. In fact, before FL, there were other similar approaches [10, 11], but FL is different from other approaches in terms of data privacy, unbalanced local data sets, and the number of local participants involved in FL [9]. In wireless applications,

TABLE 3.1
List of Abbreviations

Abbreviation	Description	Abbreviation	Description
ANN	Artificial Neural Network	MEC	Mobile Edge Computing
CNN	Convolutional Neural Network	ML	Machine Learning
CR	Cognitive Radio	MLP	Multi-Layer Perceptron
CRN	Cognitive Radio Network	NN	Neural Network
DDOS	Distributed Denial of Service	NP	Nondeterministic Polynomial
DNN	Deep Neural Network	RNN	Recurrent Neural Network
FedFMC	Fork-Merge-Consolidate	SVM	Support Vector Machine
FL	Federated Learning	SGD	Stochastic Gradient Descent
IID	Independent and Identically Distributed	TF	Tensor Flow
IoT	Internet of Thing	TFF	Tensor Flow for FL
IoV	Internet of Vehicle	UAV	Unmanned Aerial Vehicle
k-NN	k-Nearest Neighbor	WBAN	Wireless Body Area Network

FL is considered a viable technical solution for addressing growing worries about the loss of user privacy. Besides this, FL has also opened the door to plenty of new possibilities in smart medicine, finance, agriculture, and industry [12–14]. Table 3.1 shows the list of abbreviations used in this chapter.

3.2　RELATED WORK

Research describes an increasing number of wireless issues and their solutions that are based on massive data sets [15–18]. In one study, some of the data-driven research that was considered an "early adopter" utilized methodology in the communities that produced data science [17].

Other more recent work [15, 16] employs data-driven terms without addressing them directly. After carefully evaluating the studies, it is noted that the structure to address the data science challenge does not entirely fit with the techniques that have been developed and recognized. This can be understood due to the complex knowledge discovery process of data science, which requires successful recording and interpreting of the data. As a result, some of the findings' validity may be called into question.

In addition, related research provides a detailed overview of general data-mining techniques as well as case studies of how these approaches were effectively used in wireless communication areas such as the Internet of Things (IoT) and wireless sensor networks (WSN) [19–22]. Data mining is simply a small part of the knowledge extraction process from data. This demonstrates that using pre-existing data-mining methodologies is not necessarily the greatest fit for a larger problem. The selection, implementation, and evaluation of algorithms are only practicable and beneficial when the problem has been sufficiently characterized and the data, particularly its statistical properties, has been thoroughly reviewed and understood. In wireless networks, there is a lack of standardization of the approach for building models based on observed data.

Two studies [23–24] provide a comprehensive review of different intelligence algorithms such as neural networks, reinforcement learning, and swarm intelligence that has been used to address common difficulties in WSNs. Another study [25] used cognitive radio (CR) to examine many ML approaches, and yet another [20] reviewed novel ML-based algorithms for cognitive radio networks (CRN). Many research publications focus on data science approaches for the current wireless network applications [19–26].

Researchers Bulling, Blanke, and Schiele use body inertia sensors to deliver full instruction on human activity recognition (HAR) [27]. Even though their research comprises the types of data one would expect from a research paper, they focus on a specific data science topic that is the classification for a specific audience by focusing only on body sensor networks.

Due to big data applications and sophisticated models such as deep learning (DL) and ML, the model training must be disseminated across multiple machines, prompting research on decentralized machine learning [28–32]. The majority of the algorithms used in these studies are designed for machines with balanced and/or independent and identically distributed data (IID). Due to a lack of studies addressing heterogeneous and imbalanced data distribution, a rising number of researchers are interested in exploring FL [9, 36–38]. This method takes advantage of the statistical heterogeneity that arises from the fact that data are generated locally on several devices.

Researcher Wang offer techniques for FL in the context of resource-constrained edge networks [34]. While there have been some studies [39–40] that investigate using sparsification and quantization to reduce the number of reporting messages for each global iteration update, using them in FL networks is still difficult.

Multiple researchers like Bonawitz, Li, McMahan, Chen, Samarakoon, Zhang, Habachi, Park, Zeng, and Wang have investigated significant issues connected to FL implementation in wireless networks [41–51]. Researchers Bonawitz, and Li offered a detailed study of FL algorithm design and discussed different problems and their solutions for improving FL performance [41–42]. Researcher Konečný formed a couple of upgrading approaches for FL to lower uplink communication costs [43]. Researcher McMahan has demonstrated a feasible updating technique for a deep FL algorithm as well as a detailed empirical evaluation of five different FL models utilizing four different data sets [9]. The location and orientation of wireless virtual reality users are analyzed and predicted using an FL algorithm based on the echo state network [45]. Researcher Konečný suggested a new FL algorithm that reduces communication expenses [46]. Researcher Samarakoon investigated the difficulty of combining power and resource allocation for ultra-reliable low-latency communication on vehicle networks [47]. Researchers Ha and Zhang proposed a new method for minimizing the transmission and processing delays in FL algorithms [48]. Researcher Habachi employed FL algorithms to estimate traffic to maximize user data rates [49]. This earlier research was interesting; however, it assumed that wireless technology networks could simply include FL algorithms [41–49]. The FL algorithms suffered training mistakes due to the unreliability of the wireless topology and medium [50].

3.3 BACKGROUND

3.3.1 DATA SCIENCE IN WIRELESS COMMUNICATION AREAS

This section presents an overview of data science approaches and problems that occurs in wireless communication. Moreover, deep learning (DL) and federated learning (FL) models are also presented in detail.

3.3.1.1 Types of Data Science Approaches

Learning in wireless networks is a technique that provides functionalities and smartness in a range of wireless communication sectors. It has grown in popularity in recent years as a result of its success in improving network-wide performance [53], enabling intelligent behavior [54], and incorporating automation to achieve the fundamentals of self-healing and self-optimization [55]. In recent years, many learning algorithms have been utilized in multiple wireless network areas like in medium access control (MAC) [56, 57], routing [15, 16], data aggregation [31, 32], localization [33, 52], energy harvesting [35], and CR [58, 59]. Similarly, these learning algorithms are also used in multiple future Internet paradigms like mobile ad hoc networks (MANETs) [60], WSNs [24], wireless body area networks (WBAN) [61], CRN [26, 62], and cellular networks [63]. Table 3.2 outlines the learning paradigms that have previously been introduced.

- Data Mining vs. Machine Learning
  Data Mining: The goal of data mining is to find new insights in enormous data sets that have never been seen before. It aims to help humans to understand relationships in complex data, such as for businesses to group multiple customers based on their previous shopping history. Moreover, by using learning algorithms, we can extract thousands of people's purchasing patterns over time and when those results are matched with a new customer's shopping history, the trained model can automatically decide which part of the data-mining process the new customer belongs to. As a result, data mining tends to focus on fixing real-world problems through the use of ML approaches.

TABLE 3.2
Summary of Types of Learning Approaches

Learning Types	Description
Data Mining	Involves human interference to extract knowledge from a large amount of data
Machine Learning	Embodies principles of data mining, learns, and makes automatic correlations
Supervised Learning	The machine is trained using well-labeled data (inputs and outputs are known)
Unsupervised Learning	The machine is trained using unlabeled data (only inputs are known)
Semi-Supervised Learning	The machine knows some input and output pairs

Machine Learning: ML's goal is to create methods and approaches that enable computers to recognize data without being explicitly programmed. As a result, learning algorithms are commonly referred to as ML algorithms. Data miners are interested in the empirical properties of algorithms, whereas ML professionals are interested in the mathematical properties of newly developed algorithms.

- Supervised vs. Unsupervised vs. Semi-supervised

Supervised: To develop a system mode, supervised learning uses inputs and already known outcomes/outputs. The labeled training data set is a collection of both inputs and outputs that are used to train a model so that it predicts future outputs based on new inputs that were not previously added to the training set. In scenarios where there is past knowledge about the system and the data can be tagged, the usage of the supervised learning algorithm is helpful to solve wireless network issues. Multiple issues in wireless networks like MAC [56, 64–66], routing [67], link quality estimation [68, 69], WSN node clustering [70], localization [71–73], and the addition of reasoning capabilities for CR [74–80] can be solved with a supervised learning model. Supervised learning has also enhanced HAR [27, 61, 81–84], event detection [85–89], electrical charge monitoring [90, 91], security [92–94], and other wireless network applications.

Unsupervised: The opposite of supervised learning is unsupervised learning algorithms in which the model is trained on unlabeled data. The machine was given only inputs with no outputs, and learning was accomplished through data comparison. As a result, these algorithms are also appropriate for use in wireless networks. In general, problems in which no output is involved are difficult to handle. For example, unsupervised learning can automatically arrange unmanned aerial vehicles (UAVs) into groups based on their current data values and geographic location. Data aggregation [58], node pooling for WSN [95–98], data pooling [99–101], event detection [102], and various radio applications [103, 104] are all popular uses of unsupervised learning techniques in wireless networks.

Semi-supervised: Semi-supervised learning [105] is a result of combining supervised and unsupervised learning approaches in a variety of ways. When a small amount of tagged data is combined with a large amount of unlabeled data, semi-supervised learning is used. It has a lot of practical utility because it can reduce the expense of creating a completely labeled training set, especially in cases when labeling all instances isn't possible. For example, semi-supervised learning may be the ideal option for training a recognition model in gender identification systems because activities vary rapidly, certain activities remain untagged, or the gender is hesitant to engage in the data gathering process [106–108].

3.3.1.2 Problems of Data Science Approaches

As described in the previous section, data science has been successfully applied in numerous domains of wireless networks. Any wireless network challenge must first be transformed into an appropriate data mining strategy before data science approaches can be applied. In this section, we discuss some data science problems that can occur in wireless communication areas.

- Classification
  In classification, the output of the model is always in the form of discrete values or categories. The term classification refers to the notion that a given input instance's classification may be predicted. To solve classification problems, supervised learning algorithms can be utilized. These algorithms attempt to model the boundaries between sets of comparable behavior examples based on known input entities. The trained model is then used to predict future input occurrences concerning a specific class. For example, determining the application layer protocol of a traffic flow can be seen as a classification issue [17]. A few of the learning approaches that can be used to categorize data include decision trees, k-nearest neighbor (k-NN), logistic regression, support vector machine (SVM), and neural network (NN).
- Neural Network
  Artificial neural networks (ANN) or NN [109] are supervised learning algorithms inspired by brain function. They are mostly used to build non-linear sophisticated decision models, which can also be used to train regression models to predict real-valued products. At the risk of a larger processing load, NNs are known for their capacity to discover complicated non-linear correlations between input variables and identify complex trends. The NN model can be categorized into three layers: (I) input layer, (II) hidden layer, and (III) output layer. The variables in the input data correspond to the input layer while variables in the output data correspond to the output layer. Moreover, each hidden layer consists of several instances.

  Neurons use an activation function to turn input signals into output signals as they process their inputs from the previous layer. The three most popular activation functions in NN are linear, unit step, and sigmoid functions. The elements of each layer are firmly connected via links with numerical weights learned by the algorithm. With the help of hidden layer connectivity weights, the output layer creates the prediction for the provided inputs. Due to new methodologies and more powerful hardware, the algorithm has regained prominence in recent years, allowing for the creation of sophisticated models for the solution of challenging tasks. It can be said that NN is a universal approximator since they estimate any function of interest when calibrated properly [110].

  Researchers presented a NN technique in an IEEE 802.11 network based on dynamic channel selection [111]. The NN trained model is used to recognize the effects of network state on channel performance. Based on this information, the channel is dynamically selected that is anticipated to provide the optimum performance for mobile device users.
- Decision Tree
  Another supervised learning approach is known as a decision tree [112]. In the decision tree, a graph is constructed that illustrates the implications of using input values. The decision tree model is composed of three types

of nodes: (I) root or parent nodes, (II) internal or child nodes, and (III) leaf nodes. Internal nodes are sometimes also called decision nodes. Internal nodes evaluate their input to a learned expression while leaf nodes represent a class. A decision tree could be used to build simple rules that can be used to categorize future occurrences, starting with the root node till a leaf node is reached. Furthermore, in the decision tree, we achieve high accuracy only if the data are linearly separable. As a result, it is an nondeterministic polynomial (NP) complete task to construct an optimal decision tree [113].

- Logistic Regression
  The basic supervised learning technique of logistic regression [114] is extensively used to create linear classification models, which establish smooth linear decision boundaries between distinct classes. The logistic function is used to learn from models and predict future occurrences. Researcher Cerpa enhance multi-hop wireless routing by using a radio connection quality estimator [15]. Researchers Alizai and Fonseca looked at whether ML methods, such as logistic regression and NN, might outperform manually constructed estimators [115–116].

- Support Vector Machine
  Like decision trees, researcher Vapnik use the SVM approach to solve classification problems [117]. In SVM, they have divided their approach into two parts. In the first part, they have changed input data into higher-dimensional data so that it can be separated linearly. After that in the second part, they categorized the higher dimensional data to achieve multiple kernel functions.

  Different application areas are better suited for different kernel functions. The three most common kernel types are gaussian radial base, linear, and polynomial kernel functions. In CR applications, SVMs are commonly employed to conduct signal categorization. SVMs are utilized as an ML technique by researchers Hu, Huang, and Ramon to categorize signals into one of many different modulation schemes [74, 75, 118].

- K-Nearest Neighbor
  K-Nearest Neighbor (k-NN) [119] is a learning approach that uses the distance between input examples to solve classification and regression issues. It's considered a non-parametric learning algorithm since it doesn't learn a model function from training data like other supervised learning algorithms. Rather, the algorithm simply learns all past examples and then suggests the output by searching for the k closest instances in the training set. After that, it predicts the majority class among neighbors and the average output value.

  Therefore, this approach is classified as an instance-based approach. Because it is one of the simplest methods of learning, k-NN is commonly utilized. It's also known as lazy learning since a classifier isn't active until a prediction must be made, therefore no computation is done.

- Regression
  Another data-mining approach is known as regression. This approach is used to depict the value of an output variable. It belongs to the supervised

learning class that uses the optimal mathematical representation to describe a set of inputs and related outputs. Moreover, the regression approach can be categorized as a linear or non-linear function representation.

- Linear Regression
  Linear regression is used to evaluate the connection between input and output, with the outcome being a linear mixture of input variables. Linear regression is widely used in wireless networks to build a distance model for radio propagation attributes, such as a linear mathematical connection between the received signals. The strength of these signals is frequently measured in decibels.

- Non-Linear Regression
  Nonlinear regression is a regression method in which we use one or more independent variables to model observed data. For example, researcher Son used to illustrate the relationship between signal-to-noise-plus-interference ratio (SINR) and packet reception ratio (PRR), which could aid upper-layer protocol design and analysis [120].

Table 3.3 presents an overview of wireless network problems along with data science techniques (classification and regression) that can address these issues.

3.3.2 Deep Learning

To process raw data, traditional ML algorithms rely on hand-designed functions [121]. As a result, domain expertise is frequently required to build a successful ML model. In addition, for each new difficulty, the function option must be adjusted. Deep neural networks (DNNs), on the other hand, are based on representation learning, which means they can recognize and learn these qualities from raw data [122] and other sources. They frequently outperform traditional ML algorithms, particularly when there is a large amount of data. A crucial component of the brain-inspired computing paradigm DL [123], is the neural network. The working of an NN is motivated by brain neurons [124]. This model is made up of three basic layers starting with an input layer that contains input units. Next, there are multiple hidden layers that contain hidden units, and in the last layer, an output layer contains output units.

TABLE 3.3

Summary of Classification and Regression Application in Wireless Networks

	Wireless Problems					
Techniques	**Localization**	**MAC**	**Routing**	**Data Aggregation**	**Cognitive Radio**	**Relevant Studies**
Classification		✓	✓	✓	✓	[29], [9, 10], [36, 37], [97, 107]
Regression	✓		✓			[33], [94]

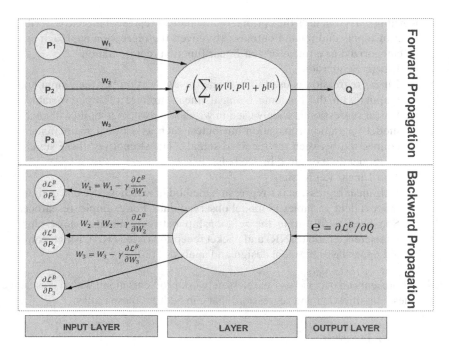

FIGURE 3.1 DNN model with forward and backward propagation.

In this model, input units contain weights that are routed through an activation function in a traditional feed-forward NN to bypass an output [125] as shown in Figure 3.1. The ReLu and Softmax functions are two activation functions [126, 127]. For example, imaging classification is done through trained DNN to generate a vector of scores called output, where the majority score represents the class to which the input image is categorized. Therefore, DNN is used to optimize the network weights so that the loss function, or the gap between the model actual output and predicted output, is as small as possible.

Before training, the data set is divided into training and test data sets. The training data set is subsequently sent into the DNN's weight optimization algorithm. The weights in DNN are calibrated using the stochastic gradient descent (SGD) model, which updates the weights using the product of (I) loss function $\mathcal{L}$ for the weight w, and (II) the learning rate γ. The formula of SGD is given in Equation 3.1.

$$W = W - \gamma \frac{\partial \mathcal{L}}{\partial W} \tag{3.1}$$

Back-propagation of the input gradient yields gradient matrices, as seen in Figure 3.1 [124]. To reduce the loss, the training iterations are repeated throughout several epochs or complete passes over the training set.

When applied to data you haven't seen before, a DNN model generalizes well by achieving high inference precision. Semi-supervised learning [128], unsupervised learning [129], and reinforcement learning [130] are examples of alternatives to supervised learning. There are also many networks and DNN architectures suited to process different types of input data, such as multi-layer perceptron (MLP) [131], convolutional neural network (CNN) [132], and recurrent neural networks (RNN) [133].

3.3.3 FEDERATED LEARNING

The concept of FL was suggested by researcher McMahan in response to data owners' privacy concerns [134]. FL is a decentralized learning technique in which several participants use their local data to train a model while being guided by a central server. In FL, each participant forwards its shared global model update to a centralized server that is used to initialize the model. This approach is different from the classical learning model in which only raw data are transmitted to the server. By doing this, we can achieve a higher privacy score without exposing participant data. Figure 3.2 shows the typical FL training process.

In FL, each participant is known as an FL participant and they work together to train a learning model which is requested by the aggregate server. The only assumption that is required for this model to work is that all involved participants must

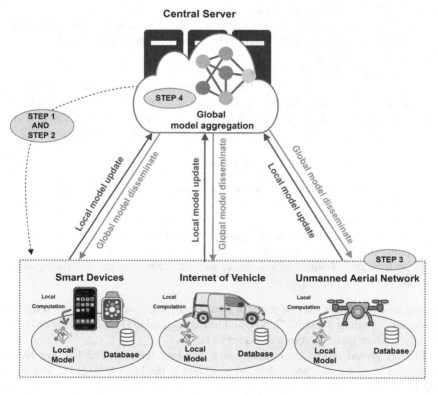

FIGURE 3.2 The typical FL training process.

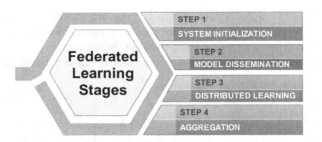

FIGURE 3.3   Stages of federated learning.

stay trustworthy meaning that they only train their private data and send their local model to the server.

The fundamental stages of FL are shown in Figure 3.3 and are briefly discussed below:

- Step 1 (System Initialization)
  The participants that will be used to train the model must be specified by the central server. The server also sets the global model and training process hyper-parameters, such as learning rate.
- Step 2 (Model Dissemination)
  After the participants have been chosen, the central server distributes the global model, which is represented by w to these participants' nodes to disseminate learning.
- Step 3 (Distributed Learning)
  Each participant updates local model parameters based on the global model w_G^k using their data and device. The current iteration number is denoted by k, and each participating node is denoted by l. The overall goal is to find an optimal parameter to minimize loss function $\mathcal{L}$ as shown in Equation 3.2.

$$w_l^{k'} = \min \mathcal{L}\left(w_l^k\right) \tag{3.2}$$

  In end, these participants forward the updated models back to the central server.
- Step 4 (Aggregation)
  The central server updates the global model by adding input from participants through the use of an FL algorithm that is designed to increase FL performance. Following that, the central server returns the modified global model to the participants.

FL training procedure may be used for a variety of ML models that rely on the SGD approach, including SVM [135], NN, and linear regression [136]. Training data in FL usually contain feature vectors and a set of labels.

FL has global model aggregation as a standard feature. The FedAvg method developed by researcher McMahan [9] is related to the local SGD [138], is a simple and traditional technique for adding local models. The FedAvg algorithm is shown in Algorithm 3.1.

Algorithm 3.1 – Federated Average Learning (FedAvg) [9]

The η clients are represented by l . E, B, and γ represents the number of epochs, minibatch size, and learning rate, respectively. ∇G is gradient of G.

[Server Side]

1:    Initialize w_0

2:    **for** each cycle k from 1 to K **do**

3:        $S_k \longleftarrow$ (random set of clients from η)

4:        **for** each client $l \in S_k$ parallely **do**

5:            $w_{k+1}^l \longleftarrow$ ClientTraining (l , w_k)

6:        **end for**

7:    $w_G^k = \frac{1}{\Sigma_{l \in \eta}} \Sigma_{l=1}^{\eta} \frac{B_l}{B} w_{l+1}^k$ (Avg Aggregation)

8:    **end for**

[ClientTraining(l , w)]

9:    $\beta \longleftarrow$ (split local dataset to minibatches of size B)

10:   **for** each local epoch k from 1 to E **do**

11:       **for** each $b \in \beta_l$ **do**

12:           $w \longleftarrow w - \gamma \nabla G(w; b)$

13:       **end for**

14:   **end for**

15:   return w to server

Table 3.4 compares FL with other learning systems. It can be seen that FL achieves high acceptable learning accuracy while maintaining privacy and reducing communication overhead.

3.3.3.1 FL Classification

Horizontal and vertical FL are the two kinds of FL that are categorized on basis of sample and feature space. Figure 3.4A shows a classification of FL.

- Horizontal FL
  It is used in situations when several data sets from distinct participants are similar in feature space but not in sample space. For example, various vehicles generated data that reflect the same feature space, namely vehicle data, but differ in the sample space like data generated from different vehicles.

TABLE 3.4

Comparison of FL with Other Learning Systems

Learning Systems	Privacy	Data Dispersal	Precision	Communication Overhead	Architecture
Centralized	Poor	Single server	High	Large	Client-Server
Distributed	Poor	Multiple server	Normal	Large	Client-Server
Peer-to-Peer	Good	Multiple clients	Normal	Large	Master-Slave
Federated learning	Good	Multiple clients and single server	Low	Small	Client-Server

- Vertical FL
  Another category of FL is vertical FL. In vertical FL, only those participants with data in the same sample space but distinct feature space exist, such as bank statements and information on a group's purchasing history [137].

3.3.3.2 FL Frameworks

Until now, multiple frameworks have been used to implement FL. The detail of these frameworks is given next.

- TensorFlow
  TensorFlow (TF) [139] is a Google-developed framework for decentralized ML and other distributed calculations. TensorFlow for FL (TFF) is made of two-layerx: (i) federated learning and (ii) FL core. The first layer is a high-level interface that allows users to do a variety of things. It integrates FL into existing TF models without the need for directly applying the FL algorithms. The second layer and TF are combined with communication operators to allow people to try out fresh and unique ideas.

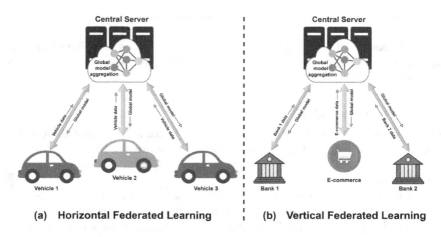

(a) **Horizontal Federated Learning**   (b) **Vertical Federated Learning**

FIGURE 3.4 FL classification (a) Horizontal Federated Learning (b) Vertical Federated Learning.

- PySyft

  PySyft [140] is a PyTorch-based framework for performing encrypted DL in untrusted contexts while preserving data, as well as implementations of related methods. PySyft is designed in such a way that it preserves the actual flow of the native Torch interface so that its operation works similarly to original PyTorch operations. A LocalTensor is automatically produced to apply the input command to Pytorch's native tensor. In PySyft, the participants are used to imitate FL. After setting all configurations of FL, the data are then distributed to the participants. The owner of the data and the storage location are then specified using a PointerTensor. Participants can also provide model changes to perform global aggregation.

- LEAF

  Federated Extended MNIST (FEMNIST) and Sentiment140 [141] data sets are examples of an open-source framework [142] that may be used as landmarks in FL. The participant is presumed to be an FL participant in these data sets, and their associated data are taken as local data kept on their equipment. Algorithms that have been freshly created are implemented using these data sets to make it possible to compare research with confidence.

3.4 APPLICATIONS OF FL

This section presents an overview of FL applications in wireless communication. The relevant studies are divided into three parts. In the first part, we discuss some studies related to the application of FL for IoT. In the second part, we discuss studies related to the application of FL for the IoV, and in the last, we discuss studies related to the application of FL in UAV. Figure 3.5 shows a typical FL architecture along with its application in wireless communication. Table 3.5 summarizes these FL applications.

3.4.1 FL FOR INTERNET OF THINGS (IoT)

To improve processing, caching, and communications in mobile edge computing (MEC), researcher Wang suggest an FL architecture [143]. Their architecture provides a path for mobile and edge devices to collaborate with FL models for system performance optimization. Researcher Liu describe an FL-based imitation learning system where, according to the researchers, sensor data are offered as a basis for robotic cloud systems with heterogeneity [144]. Moreover, they suggested that FL can increase the performance and precision of learning in a robot by exploiting the knowledge of other robots. For social recommender systems, researcher Zhou suggest an FL framework that is used to train a centralized model by collaborating with many context-aware clients and big data platforms [145].

Researcher Yin propose an IoT data cooperation architecture based on FL [146]. Moreover, they also discuss the structure of a blockchain-based method that allows different parties to collaborate in learning while maintaining anonymity. Researcher Lu work on data exchange in industrial IoT with a combination of blockchain and FL to protect anonymity [147]. FL is included in an approved blockchain's consensus process to boost processing and data interchange efficiency.

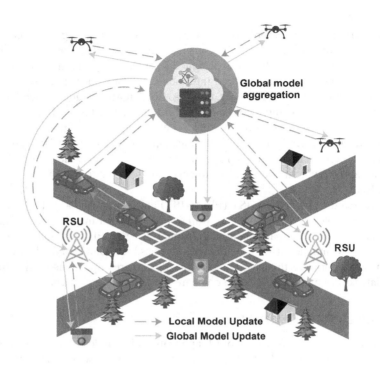

FIGURE 3.5   FL application in wireless communication.

Researcher Ren presented an FL-based architecture for big, cutting-edge IoT systems [148]. They employ FL intending to coordinate communication and informatics resources. In hybrid cloud MEC systems, researcher Fantacci employ FL to handle the problem of assigning copies of virtual machine replicas [149]. They employ FL to increase the proportion of hits as much as possible. Researcher Lu offer an asynchronous FL technique that protects MEC privacy [150]. After each cycle of client training, an asynchronous test procedure is added to assess whether client changes will be delivered to the central server. Researcher Yan presented an FL-based model for power allocation in wireless networks [151]. For local training and collaboration amongst clients, the system uses an online actor-critical algorithm in which gradients and weights are shared.

Researcher Nguyen presented an intrusion detection system based on the FL model [153]. The system effectively creates behavior profiles based on device-specific communication using FL, and no data are required for detection. Researcher Saputra utilize FL to predict electric vehicles' power demand [154]. In their scenario, charging docks act as customers that only exchange learned models with the charging provider without modifying raw user data. For cross-domain business data interchange, the researcher Verma presented a web service-based FL solution [155]. Researcher Yu proposed proactive content for edge computing based on the FL caching strategy [156]. Mobile devices act as clients in this FL paradigm, while the base station serves as the central server. Researcher Sozinov used FL to recognize

TABLE 3.5
Summary of FL Applications

Research Areas	Relevant Studies	Summary
Internet of	[143]	An MEC-based FL framework
Things	[144]	An FL-based imitation learning system
	[145]	An FL framework by collaborating with big data platforms
	[146]	A blockchain-based FL framework for IoT confidentiality
	[147]	A blockchain-based FL framework for industrial IoT
	[148]	An FL-based framework for big IoT systems
	[149]	An FL-based framework for hybrid cloud MEC systems
	[150]	An asynchronous FL framework to protect MEC privacy
	[151]	An FL-based framework for power allocation in IoT
	[153]	An FL-based intrusion detection system
	[154]	An FL-based framework to predict power demand in IoT
	[155]	A web-based FL framework for business data interchange in IoT
	[156]	An FL-based caching strategy for MEC
	[157]	An FL-based framework to recognize human activities in IoT
	[158]	An FL-based framework for multi-robot systems in IoT
	[159]	A mixed FL-based blockchain framework for privacy protection in IoT
Internet of	[160]	An asynchronous system based on blockchain FL for IoV
Vehicles	[47]	An FL-based framework to solve issues of shared power and allocation of resources
	[162]	An FL-based framework for image classification in IoV
	[163]	An FL-based framework for resource sharing in IoV
Unmanned	[152]	An FL-based approach to detect the jamming attack in UAV
Aerial Vehicle	[164]	An FL-based MFG framework for UAV
	[181]	An FL-based framework to investigate the visible light communication in UAV
	[182]	An incentive-based FL framework for UAV

human activities [157]. FL might attain accuracy equivalent to centralized learning, according to the researchers. For multi-robot systems, researcher Zhou suggest an FL-based real-time data processing architecture [158]. To determine the relevance of data and preserve it in a decentralized manner, researcher Doku employ a mix of FL and blockchain [159].

3.4.2 FL FOR INTERNET OF VEHICLES (IoV)

Multiple studies have been also presented on the usage of FL in the Internet of Vehicles (IoV) scenarios. Researcher Zhang discussed difference challenges to secure data interchange in FL [160]. They propose an asynchronous system based on blockchain FL for IoV. Concerns about privacy and the burden of communication lead the researcher Samarakoon examine the issue of shared power and allocation of resources for ultra-reliable, low-latency communication in a vehicular environment [47]. The network-wide queue lengths that represent the distribution network state are estimated using FL. Researcher Ye discussed Image classification in vehicular IoT using FL in [162]. Local image quality, as well as the processing power of each vehicle, is considered by the aggregate server to pick local calculated models of vehicles.

Researcher Lu presented the FL approach for resource sharing in IoT-based vehicular networks [163]. In their approach, they work on protecting the privacy of local updates by introducing a differential-based privacy scheme. They asynchronously use FL so that each vehicle shares its updated model in a peer-to-peer way rather than focusing on a decentralized way.

3.4.3 FL for Unmanned Aerial Vehicle (UAV)

For mission-critical operations like fighting fires or covering disaster locations, many UAVs might sometimes fly from one location to another.

Researcher Mowla presented an FL-based architecture for detection of jamming attack in unmanned aerial vehicles [152]. They utilize Dempster-Shafer theory to increase the FL architecture productivity.

UAV control becomes extremely difficult because of massive collapses that occurred because of the wind and other uncontrollable circumstances. To keep communications between UAVs to a minimum, the mean-field game (MFG) is utilized by researcher Shiri for enormous control of UAVs by solving a couple stochastic differential equations [164]. Researcher Lim presented an incentive-based scheme through which multiple UAVs collaborate with FL servers to improve system performance [182]. Researcher Wang focuses on constructing a convolutional auto-encoder for light distribution in UAV communications using FL [181].

3.5 RESEARCH CHALLENGES AND FUTURE DIRECTIONS

In this section, multiple research challenges and future directions of FL are discussed.

3.5.1 Research Challenges

This section deals with the research challenges of FL. FL research challenges are divided into three types: (I) security and privacy, (II) data distribution, and (III) algorithm related.

3.5.1.1 Security and Privacy

FL's major goal is to protect user privacy by requiring user to only provide trained model attributes rather than sending actual data. But a recent study has found that when an FL participant device or FL servers are malicious, privacy and safety problems may develop [161]. This is particularly incompatible with FL's goal, as the resulting global model may be corrupted, and participant privacy may be compromised during model training.

Researcher Xu presented an FL framework named as VerifyNet. It uses a double-masking protocol to improve the privacy of local gradients [165]. In addition, VerifyNet delivers a method for ensuring the integrity of aggregated data to get results from the central server, and function technologies are demonstrated using real-world data. Researcher Nasr proposed a white-box approach for inference [166]. FL models will be subjected to a privacy examination. Researcher Wang investigated FL model privacy risk and GAN-based attacks [167]. This allows servers to target and corrupt the privacy of a single client. Researcher Triastcyn used Bayesian differential privacy to provide greater privacy assurance for the clients [168].

Researcher Zhang developed an FL approach by focusing on client privacy [169]. This approach uses a homomorphic cryptosystem that can be applied to encrypted data without disclosing the data's values. To create distributed encryptions and minimize communication costs, a distributed selective SGD algorithm was used. To validate the clients, an authentication approach was also incorporated. Researcher Sharma presented a study in which earlier FL models in hostile environments are improved in terms of their security [170].

Researcher Cao discussed poisoning attack outcomes and performance containing infected data along with the number of attackers and proposed a strategy for eliminating infected local models during global model training [171]. Researcher Gao presented a privacy protection-based architecture for heterogeneous FL, which used a secure multi-party learning technique from start to finish [172]. Researcher Bonawitz presented a safe aggregation approach in [173], which enables clients to encrypt their local models while enabling the FL server to combine them without decryption However, analyzing the global aggregated model can assist in revealing the activity of specific clients. As a result, rather than maintaining collected data, developing methods that provide participant privacy is crucial.

3.5.1.2 Data Distribution

In FL, the data are normally divided into two parts. The first part is known as IID and the second is known as non-IID. Unbalanced quantity, features, and labels can all lead to non-IID data. In a real-world scenario, non-IID data are far more widespread. Researcher Zhao showed that non-IID circumstances can cause a considerable reduction in FL model performance, which is driven by weight divergence due to device, class, and population dispersion [174]. They also advised creating a global common data set split to aid with non-IID data training.

They discovered that sharing just 5% of data can boost 30% inaccuracy. This, however, raises model communication costs and presupposes that such data set partitioning is always available.

Reserracher Kopparapu introduced a Federated Fork-Merge-Consolidate (FedFMC) method for training non-IID data [175]. In this approach, the global model was split into multiple sub-models, and splitting was done based on how a sub-model performs on each device data set. Devices were grouped to create separate global models for each data set. As a result, depending on their data set qualities, the grouped devices might focus on different portions of the model. All the models are eventually integrated. However, by grouping devices with similar properties, this strategy renders the model more vulnerable to an adversarial assault.

3.5.1.3 Related to Algorithm

One of the major challenges in FL, as with practically any decentralized algorithm, is the method's convergence under constrained communication and processing resources. Researcher Wang discussed the performance limits of gradient descent-based FL for convex loss functions [34]. This non-convex loss function-based approach is relevant to DNNs as DNNs are also trained using non-convex functions. In addition, application-specific factors are also important to investigate. These application-specific factors can be in the form of an optimal number of participants, the clustering of local trainees, and the recurrence of local updates and global aggregation. Moreover, in models like federated DNNs, even a small number of updates can be crucial to perform at low-power sensor nodes in IoT.

3.5.2 FUTURE DIRECTIONS

Besides all the previously mentioned issues, there are still hurdles and new research directions to be considered in applying FL on a large scale. Some of the future directions are given next:

3.5.2.1 Resource and Communication

To obtain output accuracy from the FL, a series of communications between the participant device and the FL server is required. Each update for large DL models, for example, CNN can contain millions of parameters [176]. Because of the high complexity of the updates, multiple issues are generated like training delays and significant communication costs. Furthermore, the bottleneck may get worse because of the unpredictable network conditions of participating devices [177] and asymmetry in Internet connection speeds, with the upload speed being quicker than the download speed [36]. Due to this, it is important to increase FL efficiency. Moreover, FL comprises the participation of heterogeneous devices with varying data sets, processing capabilities, energy levels, and participation willingness. Given the variety of devices and resource limits, resource allocation must be optimized to achieve high training performance.

3.5.2.2 Power and Data

Researcher Ye suggested a technique in which each edge device is compensated for power and data quality computations [162]. On the other hand, they also discussed a basic FL framework with a single server. Multiple access points as servers in FL are possible in a practical vehicular network situation, adding to the complexity of the reputation management challenge. As a result, further study into multi-server FL designs is required for their use in in-vehicle networks.

3.5.2.3 Distributed Vulnerabilities

As FL is a new emerging paradigm, it is open to various communication security vulnerabilities like distributed denial-of-service (DDoS) [178] and jammer attacks [179]. In a jamming attack, an attacker blocks the communication signals by interfering with devices and the server. Because of this type of attack, system performance downgrades along with model upload/download mistakes. Anti-jamming approaches [180] such as frequency hopping, which involves sending a second copy of the model update over multiple frequencies, can be used to combat the problem.

3.6 CONCLUSION

In this article, we highlighted the expanding potential and significance of the data-driven research paradigm, along with FL in wireless applications. FL is emerging as a new decentralized learning paradigm that overcomes wireless network connection costs as well as data privacy by doing model training in a distributed manner. We began by providing a general overview of data-driven techniques along with DNN. After that, we talked about FL and its basics. We then enlist many wireless network applications for FL utilization, ranging from IoT, IoV, to UAV. We also highlighted the importance of FL in tackling some of the challenges, primarily those relating to security and privacy, data distribution, and algorithm. Finally, we talked about potential research possibilities for using FL on a larger scale. We expect that this chapter will generate additional interest in this new area and inspire additional research efforts toward the full realization of FL-based wireless communication.

REFERENCES

1. V. Dhar, "Data science and prediction," Commun. ACM, vol. 56, pp. 64–73, 2013.
2. S. H. Liao, P.H. Chu, P.Y. Hsiao, "Data mining techniques and applications—A decade review from 2000 to 2011," Expert Syst. Appl, vol. 39, pp. 11303–11311, 2012.
3. S. Palaniappan, R. Awang, "Intelligent heart disease prediction system using data mining techniques," in Proceedings of the 2008 IEEE/ACS International Conference on International Conference on Computer Systems and Applications, Doha, Qatar, 31 March–4 April 2008, pp. 108–115.
4. M.A. Russell, Mining the Social Web: Data Mining Facebook, Twitter, LinkedIn, Google+, GitHub, and More; O'Reilly Media, Inc.: Sebastopol, CA, USA, 2013.
5. G. Adomavicius, A. Tuzhilin, "Using data mining methods to build customer profiles," Computer, vol. 34, pp. 74–82, 2001.
6. H. Chen, A. Turk, S.S. Duri, C. Isci, A.K. Coskun, "Automated system change discovery and management in the cloud," IBM J. Res. Dev., vol. 60, no. 2, pp. 1–2:10, 2016.
7. E. Kartsakli, A. Antonopoulos, L. Alonso, C. Verikoukis, "A cloud-assisted random linear network coding medium access control protocol for healthcare applications," Sensors, vol. 14, pp. 4806–4830, 2014.
8. E. Bastug, M. Bennis, E. Zeydan, M. A. Kader, A. Karatepe, A.S. Er, M. Debbah, "Big data meets telcos: A proactive caching perspective," J. Commun. Netw., vol. 17, pp. 549–558, 2015.
9. B. McMahan, E. Moore, D. Ramage, S. Hampson, and B. A. y Arcas, "Communication-efficient learning of deep networks from decentralized data," in Artificial Intelligence and Statistics, Apr. 2017, pp. 1273–1282.
10. D. Povey, X. Zhang, and S. Khudanpur, "Parallel training of deep neural networks with natural gradient and parameter averaging," 2015, arXiv:1410.7455. [Online]. Available: https://arxiv.org/pdf/1410.7455.pdf
11. N. Neverova et al., "Learning human identity from motion patterns," IEEE Access, vol. 4, pp. 1810–1820, Apr. 2016.

12. N.H. Tran, W. Bao, A. Zomaya, N.M. NH, and C.S. Hong, "Federated learning over wireless networks: Optimization model design and analysis," in Proc. IEEE INFOCOM, Paris, France, Apr. 2019, pp. 1387–1395.

13. S. Samarakoon, M. Bennis, W. Saad, and M. Debbah, "Federated learning for ultra-reliable low-latency V2V communications," in Proc. IEEE Global Commun. Conf., Abu Dhabi, United Arab Emirates, Dec. 2018, pp. 1–7.

14. J. Lee, J. Sun, F. Wang, S. Wang, C.-H. Jun, and X. Jiang, "Privacy preserving patient similarity learning in a federated environment: Development and analysis," JMIR Med. Informat., vol. 6, no. 2, pp. e20, 2018.

15. T. Liu, A.E. Cerpa, Foresee (4C): Wireless link prediction using link features. in Proceedings of the 2011 10th International Conference on Information Processing in Sensor Networks (IPSN), Chicago, IL, USA, 12–14 April 2011, pp. 294–305.

16. T. Liu, A. E. Cerpa, "Temporal adaptive link quality prediction with online learning," ACM Trans. Sen. Netw., vol. 10, pp. 1–41, 2014.

17. M. Crotti, M. Dusi, F. Gringoli, L. Salgarelli, "Traffic classification through simple statistical fingerprinting," ACM SIGCOMM Comput. Commun. Rev., vol. 37, pp. 5–16, 2007.

18. S.V. Radhakrishnan, A.S. Uluagac, R. Beyah, "GTID: A technique for physical device and device type fingerprinting," IEEE Trans. Depend. Secur. Comput., vol. 12, pp. 519–532, 2014.

19. C.W. Tsai, C.F. Lai, M.C. Chiang, L.T. Yang, "Data mining for Internet of Things: A survey," IEEE Commun. Surv. Tutor., vol. 16, pp. 77–97, 2014.

20. A. Mahmood, K. Shi, S. Khatoon, M. Xiao, "Data mining techniques for wireless sensor networks: A survey," Int. J. Distrib. Sens. Network, vol. 2013, 2013, doi:10.1155/2013/406316.

21. F. Chen, P. Deng, J. Wan, D. Zhang, A.V. Vasilakos, X. Rong, "Data mining for the Internet of Things: Literature review and challenges," Int. J. Distrib. Sens. Network, vol. 501, 2015, doi:10.1155/2015/431047.

22. M. Di, E.M. Joo, "A survey of machine learning in wireless sensor networks from networking and application perspectives," in Proceedings of the 2007 6th International Conference on IEEE Information, Communications & Signal Processing, Singapore, 10–13 December 2007, pp. 1–5.

23. M. Abu Alsheikh, S. Lin, D. Niyato, H.P. Tan, "Machine learning in wireless sensor networks: Algorithms, strategies, and applications," IEEE Commun. Surv. Tutor., vol. 16, no. 4, pp. 1996–2018, 2014.

24. R.V. Kulkarni, A. Förster, G.K. Venayagamoorthy, "Computational intelligence in wireless sensor networks: A survey," IEEE Commun. Surv. Tutor., vol. 13, pp. 68–96, 2011.

25. M. Bkassiny, Y. Li, S.K. Jayaweera, "A survey on machine-learning techniques in cognitive radios," IEEE Commun. Surv. Tutor., vol. 15, pp. 1136–1159, 2013.

26. K.M. Thilina, K.W. Choi, N. Saquib, E. Hossain, "Machine learning techniques for cooperative spectrum sensing in cognitive radio networks," IEEE J. Sel. Areas Commun., vol. 31, pp. 2209–2221, 2013.

27. A. Bulling, U. Blanke, B. Schiele, "A tutorial on human activity recognition using body-worn inertial sensors," ACM Comput. Surv. (CSUR), vol. 46, pp. 33, 2014.

28. C. Ma et al., "Distributed optimization with arbitrary local solvers," Optimization Methods and Software, vol. 32, no. 4, pp. 813–848, Jul. 2017.

29. O. Shamir, N. Srebro, and T. Zhang, "Communication-efficient distributed optimization using an approximate newton-type method," in ICML, Beijing, China, 2014, pp. II–1000–II–1008.

30. J. Wang and G. Joshi, "Cooperative SGD: A unified Framework for the Design and Analysis of Communication-Efficient SGD Algorithms," arXiv:1808.07576 [cs, stat], Jan. 2019.

31. S. U. Stich, Local SGD Converges Fast and Communicates Little. arXiv, 2018. doi: 10.48550/ARXIV.1805.09767.
32. F. Zhou and G. Cong, "On the convergence properties of a K-step averaging stochastic gradient descent algorithm for nonconvex optimization," in Proceedings of the 27th International Joint Conference on Artificial Intelligence, Stockholm, Sweden, Jul. 2018, pp. 3219–3227.
33. Vanheel, F.; Verhaevert, J.; Laermans, E.; Moerman, I.; Demeester, P. Automated linear regression tools improve rssi wsn localization in multipath indoor environment. EURASIP J. Wirel. Commun. Netw. 2011, 2011, 1–27
34. S. Wang et al., "Adaptive federated learning in resource constrained edge computing systems," IEEE J. Sel. Areas Commun., vol. 37, no. 6, pp. 1205–1221, Jun. 2019.
35. T. Li et al., "Federated optimization for heterogeneous networks," in Proceedings of the 1st Adaptive & Multitask Learning, ICML Workshop, 2019, Long Beach, CA, p. 16.
36. J. Konečný et al., "Federated Learning: Strategies for Improving Communication Efficiency," http://arxiv.org/abs/1610.05492, Oct. 2016.
37. V. Smith, C.-K. Chiang, M. Sanjabi, and A. Talwalkar, "Federated Multitask Learning," in NeurIPS'17, CA, U.S., 2017, pp. 4427–4437.
38. C.T. Dinh et al., "Federated learning with proximal stochastic variance reduced gradient algorithms," in 49th International Conference on Parallel Processing - ICPP, Edmonton Canada, Aug. 2020, pp. 1–11.
39. H. Wang et al., "ATOMO: Communication-efficient learning via atomic sparsification," in NeurIPS, 2018, pp. 9850–9861.
40. H. Zhang et al., "ZipML: Training linear models with end-to-end low precision, and a little bit of deep learning," in International Conference on Machine Learning, Jul. 2017, pp. 4035–4043.
41. K. Bonawitz, H. Eichner, W. Grieskamp, D. Huba, A. Ingerman, V. Ivanov, C. M. Kiddon, J. Konečný, S. Mazzocchi, B. McMahan, T. V. Overveldt, D. Petrou, D. Ramage, and J. Roselander, "Towards federated learning at scale: System design," in Proc. Systems and Machine Learning Conference, Stanford, CA, USA, 2019.
42. T. Li, A.K. Sahu, A. Talwalkar, and V. Smith, "Federated learning: Challenges, methods, and future directions," IEEE Signal Process. Mag., vol. 37, no. 3, pp. 50–60, May 2020.
43. J. Konečný, H.B. McMahan, D. Ramage, and P. Richtárik, "Federated optimization: Distributed machine learning for on-device intelligence," arXiv preprint arXiv:1610.02527, Oct. 2016.
45. M. Chen, O. Semiari, W. Saad, X. Liu, and C. Yin, "Federated echo state learning for minimizing breaks in presence in wireless virtual reality networks," IEEE Trans. Wireless Commun., vol. 19, no. 1, pp. 177–191, Jan. 2020.
46. J. Konečný, B. McMahan, and D. Ramage, "Federated optimization: Distributed optimization beyond the datacenter," arXiv preprint arXiv:1511.03575, Nov. 2015.
47. S. Samarakoon, M. Bennis, W. Saad, and M. Debbah, "Distributed federated learning for ultra-reliable low-latency vehicular communications," IEEE Trans. Commun., vol. 68, no. 2, pp. 1146–1159, Feb. 2020.
48. S. Ha, J. Zhang, O. Simeone, and J. Kang, "Coded federated computing in wireless networks with straggling devices and imperfect CSI," in Proc. IEEE Int. Symp. Inf. Theory (ISIT), Paris, France, Jul. 2019.
49. O. Habachi, M.A. Adjif, and J.P. Cances, "Fast uplink grant for NOMA: A federated learning-based approach," arXiv preprint arXiv:1904.07975, Mar. 2019.
50. J. Park, S. Samarakoon, M. Bennis, and M. Debbah, "Wireless network intelligence at the edge," Proc. IEEE, vol. 107, no. 11, pp. 2204–2239, Nov. 2019.
51. Q. Zeng, Y. Du, K. Huang, and K.K. Leung, "Energy-efficient radio resource allocation for federated edge learning," in Proc. IEEE Int. Conf. Commun. Workshop, Dublin, Ireland, Jun. 2020.

52. Tennina, S.; Di Renzo, M.; Kartsakli, E.; Graziosi, F.; Lalos, A.S.; Antonopoulos, A.; Mekikis, P.V.; Alonso, L. WSN4QoL: A WSN-oriented healthcare system architecture. Int. J. Distrib. Sens. Netw. 2014, 2014, doi:10.1155/2014/503417

53. K.L.A. Yau, P. Komisarczuk, P.D. Teal, "Reinforcement learning for context awareness and intelligence in wireless networks: Review, new features and open issues," J. Netw. Comput. Appl., vol. 35, pp. 253–267, 2012.

54. G.K.K. Venayagamoorthy, "A successful interdisciplinary course on computational intelligence," IEEE Comput. Intell. Mag., vol. 4, pp. 14–23, 2009.

55. E.J. Khatib, R. Barco, P. Munoz, L. Bandera, I. De, I. Serrano, "Self-healing in mobile networks with big data," IEEE Commun. Mag., vol. 54, pp. 114–120, 2016.

56. M. Sha, R. Dor, G. Hackmann, C. Lu, T.S. Kim, T. Park, "Self-adapting mac layer for wireless sensor networks," in Proceedings of the 2013 IEEE 34th Real-Time Systems Symposium (RTSS), Vancouver, BC, Canada, 3–6 December 2013, pp. 192–201.

57. V. Esteves, A. Antonopoulos, E. Kartsakli, M. Puig-Vidal, P. Miribel-Català, C. Verikoukis, "Cooperative energy harvesting-adaptive MAC protocol for WBANs," Sensors, vol. 15, pp. 12635–12650, 2015.

58. S. Yoon, C. Shahabi, "The Clustered AGgregation (CAG) technique leveraging spatial and temporal correlations in wireless sensor networks," ACM Trans. Sens. Netw. (TOSN), vol. 3, 2007. doi:10.1145/1210669.1210672.

59. X. Chen, X. Xu, J.Z. Huang, Y. Ye, "TW-k-means: Automated two-level variable weighting clustering algorithm for multiview data," IEEE Trans. Knowl. Data Eng., vol. 25, pp. 932–944, 2013.

60. A. Förster, "Machine learning techniques applied to wireless ad-hoc networks: Guide and survey," in Proceedings of the 3rd International Conference on ISSNIP 2007 IEEE Intelligent Sensors, Sensor Networks and Information, Melbourne, Australia, 3–6 December 2007, pp. 365–370.

61. O.D. Lara, M.A. Labrador, "A survey on human activity recognition using wearable sensors," IEEE Commun. Surv. Tutor., vol. 15, pp. 1192–1209, 2013.

62. C. Clancy, J. Hecker, E. Stuntebeck, T.O. Shea, "Applications of machine learning to cognitive radio networks," IEEE Wirel. Commun., vol. 14, pp. 47–52, 2007.

63. T. Anagnostopoulos, C. Anagnostopoulos, S. Hadjiefthymiades, M. Kyriakakos, A. Kalousis, "Predicting the location of mobile users: A machine learning approach," in Proceedings of the 2009 International Conference on Pervasive Services, New York, NY, USA, 13 July 2009, pp. 65–72.

64. R.V. Kulkarni, G.K. Venayagamoorthy, "Neural network based secure media access control protocol for wireless sensor networks," in Proceedings of the 2009 International Joint Conference on Neural Networks, Atlanta, GA, USA, 14–19 June 2009, pp. 1680–1687.

65. Kim, M.H.; Park, M.G. Bayesian statistical modeling of system energy saving effectiveness for MAC protocols of wireless sensor networks. In Software Engineering, Artificial Intelligence, Networking and Parallel/Distributed Computing; Springer: Heidelberg, Germany, 2009; pp. 233–245.

66. Y.J. Shen, M.S. Wang, "Broadcast scheduling in wireless sensor networks using fuzzy Hopfield neural network," Expert Syst. Appl., vol. 34, pp. 900–907, 2008.

67. J. Barbancho, C. León, J. Molina, A. Barbancho, "Giving neurons to sensors. QoS management in wireless sensors networks," in Proceedings of the IEEE Conference on Emerging Technologies and Factory Automation (2006 ETFA'06), Prague, Czech Republic, 20–22 September, 2006, pp. 594–597.

68. T. Liu, A.E. Cerpa, "Data-driven link quality prediction using link features," ACM Trans. Sens. Netw. (TOSN), vol. 10, 2014, doi:10.1145/2530535.

69. Y. Wang, M. Martonosi, L.S. Peh, "Predicting link quality using supervised learning in wireless sensor networks," ACM SIGMOBILE Mobile Comput. Commun. Rev., vol. 11, pp. 71–83, 2007.
70. G. Ahmed, N.M. Khan, Z. Khalid, R. Ramer, "Cluster head selection using decision trees for wireless sensor networks," in Proceedings of the ISSNIP 2008 International Conference on Intelligent Sensors, Sensor Networks and Information Processing, Sydney, Australia, 15–18 December 2008. pp. 173–178.
71. A. Shareef, Y. Zhu, M. Musavi, "Localization using neural networks in wireless sensor networks," in Proceedings of the 1st International Conference on MOBILeWireless MiddleWARE, Operating Systems, and Applications, ICST, Brussels, Belgium, 13 February 2008.
72. S.H. Chagas, J.B. Martins, L.L. De Oliveira, "An approach to localization scheme of wireless sensor networks based on artificial neural networks and genetic algorithms," in Proceedings of the 2012 IEEE 10th International New Circuits and systems Conference (NEWCAS), Montreal, QC, Canada, 2012, pp. 137–140.
73. D.A. Tran, T. Nguyen, "Localization in wireless sensor networks based on support vector machines," IEEE Trans. Parallel Distrib. Syst., vol. 19, pp. 981–994, 2008.
74. H. Hu, J. Song, Y. Wang, "Signal classification based on spectral correlation analysis and SVM in cognitive radio," in Proceedings of 22nd International Conference on the Advanced Information Networking and Applications (AINA 2008), Okinawa, Japan, 25–28 March 2008, pp. 883–887.
75. Y. Huang, H. Jiang, H. Hu, Y. Yao, "Design of learning engine based on support vector machine in cognitive radio," in Proceedings of the International Conference on IEEE Computational Intelligence and Software Engineering (CiSE 2009), Wuhan, China, 11–13 December 2009, pp. 1–4.
76. V.K. Tumuluru, P. Wang, D. Niyato, "A neural network-based spectrum prediction scheme for cognitive radio," in Proceedings of the 2010 IEEE International Conference on IEEE Communications (ICC), Cape Town, South Africa, 23–27 May 2010, pp. 1–5.
77. N. Baldo, M. Zorzi, "Learning and adaptation in cognitive radios using neural networks," in Proceedings of the 2008 5th IEEE Consumer Communications and Networking Conference, Las Vegas, NV, USA, 10–12 January 2008, pp. 998–1003.
78. Y.J. Tang, Q.Y. Zhang, W. Lin, "Artificial neural network-based spectrum sensing method for cognitive radio," in Proceedings of the 2010 6th International Conference on Wireless Communications Networking and Mobile Computing (WiCOM), Chengdu, China, 23–25 September 2010, pp. 1–4.
79. G. Xu, Y. Lu, "Channel and modulation selection based on support vector machines for cognitive radio," in Proceedings of the 2006 International Conference on Wireless Communications, Networking and Mobile Computing, Wuhan, China, 22–24 September 2006, pp. 1–4.
80. M. Petrova, P. Mähönen, A. Osuna, "Multi-class classification of analog and digital signals in cognitive radios using support vector machines," in Proceedings of the 2010 7th International Symposium on Wireless Communication Systems (ISWCS), York, UK, 19–22 September 2010, pp. 986–990.
81. A. Mannini, A.M. Sabatini, "Machine learning methods for classifying human physical activity from on-body accelerometers," Sensors, vol. 10, pp. 1154–1175, 2010.
82. J.H. Hong, N.J. Kim, E.J. Cha, T.S. Lee, "Classification technique of human motion context based on wireless sensor network," in Proceedings of the 27th Annual International Conference of the Engineering in Medicine and Biology Society (IEEE-EMBS 2005), Shanghai, China, 17–18 January 2006, pp. 5201–5202.
83. L. Bao, S.S. Intille, Activity Recognition from User-Annotated Acceleration Data. In Pervasive Computing; Springer: Heidelberg, Germany; 2004; pp. 1–17.

84. A. Bulling, J.A. Ward, H. Gellersen, "Multimodal recognition of reading activity in transit using body-worn sensors," ACM Trans. Appl. Percept., vol. 9, 2012, doi:10.1145/2134203.2134205.

85. L. Yu, N. Wang, X. Meng, "Real-time forest fire detection with wireless sensor networks," in Proceedings of the 2005 International Conference on Wireless Communications, Networking and Mobile Computing, Wuhan, China; 23–26 September 2005, pp. 1214–1217.

86. M. Bahrepour, N. Meratnia, P.J. Havinga, "Use of AI techniques for residential fire detection in wireless sensor networks," in Proceedings of the AIAI 2009Workshop Proceedings, Thessaloniki, Greece, 23–25 April 2009, pp. 311–321.

87. M. Bahrepour, N. Meratnia, M. Poel, Z. Taghikhaki, P.J. Havinga, "Distributed event detection in wireless sensor networks for disaster management," in Proceedings of the 2010 2nd International Conference on Intelligent Networking and Collaborative Systems (INCOS), Thessaloniki, Greece, 24–26 November 2010, pp. 507–512.

88. A. Zoha, A. Imran, A. Abu-Dayya, A. Saeed, "A machine learning framework for detection of sleeping cells in LTE network," in Proceedings of the Machine Learning and Data Analysis Symposium, Doha, Qatar, 3–4 March 2014.

89. R.M. Khanafer, B. Solana, J. Triola, R. Barco, L. Moltsen, Z. Altman, P. Lazaro, "Automated diagnosis for UMTS networks using Bayesian network approach," IEEE Trans. Veh. Technol., vol. 57, pp. 2451–2461, 2008.

90. A. Ridi, C. Gisler, J. Hennebert, "A survey on intrusive load monitoring for appliance recognition," in Proceedings of the 2014 22nd International Conference on Pattern Recognition (ICPR), Stockholm, Sweden, 24–28 August 2014, pp. 3702–3707.

91. H.H. Chang, H.T. Yang, C.L. Lin, Load identification in neural networks for a non-intrusive monitoring of industrial electrical loads. In Computer Supported Cooperative Work in Design IV; Springer: Melbourne, Australia, 2007; pp. 664–674.

92. J.W. Branch, C. Giannella, B. Szymanski, R. Wolff, H. Kargupta, "In-network outlier detection in the wireless sensor networks," Knowl. Inf. Syst., vol. 34, pp. 23–54, 2013.

93. S. Kaplantzis, A. Shilton, N. Mani, Y.A. Sekercioglu, "Detecting selective forwarding attacks in wireless sensor networks using support vector machines," in Proceedings of the 2007 3rd International Conference on IEEE Intelligent Sensors, Sensor Networks and Information (2007 ISSNIP), Melbourne, Australia, 3–6 December 2007, pp. 335–340.

94. R.V. Kulkarni, G.K. Venayagamoorthy, A.V. Thakur, S.K. Madria, "Generalized neuron based secure media access control protocol for wireless sensor networks," in Proceedings of IEEE Symposium on Computational Intelligence in Multi-Criteria Decision-Making (MCDM'09), Nashville, TN, USA, 30 March–2 April 2009, pp. 16–22.

95. Yoon, S.; Shahabi, C. The Clustered AGgregation (CAG) technique leveraging spatial and temporal correlations in wireless sensor networks. ACM Trans. Sens. Netw. (TOSN) 2007, 3, doi:10.1145/1210669.1210672.

96. H. He, Z. Zhu, E. Mäkinen, "A neural network model to minimize the connected dominating set for self-configuration of wireless sensor networks," IEEE Trans. Neural Network, vol. 20, pp. 973–982, 2009.

97. T. Kanungo, D.M. Mount, N.S. Netanyahu, C.D. Piatko, R. Silverman, A.Y. Wu, "An efficient k-means clustering algorithm: Analysis and implementation," IEEE Trans. Pattern Anal. Mach. Intell., vol. 24, pp. 881–892, 2002.

98. C. Liu, K. Wu, J. Pei, "A dynamic clustering and scheduling approach to energy saving in data collection from wireless sensor networks," in Proceedings of the IEEE SECON, Santa Clara, CA, USA, 26–29 September 2005, pp. 374–385.

99. A. Taherkordi, R. Mohammadi, F. Eliassen, "A communication-efficient distributed clustering algorithm for sensor networks," in Proceedings of the 22nd International Conference on Advanced Information Networking and Applications-Workshops (AINAW 2008), Okinawa, Japan, 25–28 March 2008, pp. 634–638.

100. L. Guo, C. Ai, X. Wang, Z. Cai, Y. Li, "Real time clustering of sensory data in wireless sensor networks," in Proceedings of the 2009 IEEE 28th International Performance Computing and Communications Conference (IPCCC), Scottsdale, AZ, USA, 14–16 December 2009, pp. 33–40.
101. K. Wang, S.A. Ayyash, T.D. Little, P. Basu, "Attribute-based clustering for information dissemination in wireless sensor networks," in Proceedings of the 2nd Annual IEEE Communications Society Conference on Sensor and ad hoc Communications and Networks (SECON'05), Santa Clara, CA, USA, 26–29 September 2005, pp. 498–509.
102. Y. Ma, M. Peng, W. Xue, X. Ji, "A dynamic affinity propagation clustering algorithm for cell outage detection in self-healing networks," in Proceedings of the 2013 IEEE Wireless Communications and Networking Conference (WCNC), Shanghai, China, 7–10 April 2013, pp. 2266–2270.
103. T.C. Clancy, A. Khawar, T.R. Newman, "Robust signal classification using unsupervised learning," IEEE Trans. Wireless. Commun., vol. 10, pp. 1289–1299, 2011.
104. N. Shetty, S. Pollin, P. Pawełczak, "Identifying spectrum usage by unknown systems using experiments in machine learning," in Proceedings of the 2009 IEEE Wireless Communications and Networking Conference, Budapest, Hungary, 5–8 April 2009, pp. 1–6.
105. I.H. Witten, E. Frank, Data Mining: Practical Machine Learning Tools and Techniques, 2nd ed. (Morgan Kaufmann Series in Data Management Systems); Morgan Kaufmann Publishers Inc.: San Francisco, CA, USA, 2005.
106. D. Guan, W. Yuan, Y.K. Lee, A. Gavrilov, S. Lee, "Activity recognition based on semi-supervised learning," in Proceedings of the 13th IEEE International Conference on Embedded and Real-Time Computing Systems and Applications, Daegu, Korea, 21–24, August 2007, pp. 469–475.
107. M. Stikic, D. Larlus, S. Ebert, B. Schiele, "Weakly supervised recognition of daily life activities with wearable sensors," IEEE Trans. Pattern Anal. Mach. Intell., vol. 33, pp. 2521–2537, 2011.
108. T. Huynh, B. Schiele, "Towards less supervision in activity recognition from wearable sensors," in Proceedings of the 2006 10th IEEE International Symposium on Wearable Computers, Montreux, Switzerland, 11–14 October 2006; pp. 3–10.
109. S.S. Haykin, Neural Networks and Learning Machines; Pearson Education: Upper Saddle River, NJ, USA, 2009.
110. K. Hornik, M. Stinchcombe, H. White, "Multilayer feedforward networks are universal approximators," Neural Networks, vol. 2, pp. 359–366, 1989.
111. N. Baldo, B.R. Tamma, B. Manoj, R. Rao, M. Zorzi, "A neural network based cognitive controller for dynamic channel selection," in Proceedings of the IEEE International Conference on Communications (ICC'09), Dresden, Germany, 14–18 June 2009, pp. 1–5.
112. L. Rokach, O. Maimon, Data Mining with Decision Trees: Theory and Applications; World Scientific Publishing Company: Singapore, 2008.
113. S.R. Safavian, D. Landgrebe, "A survey of decision tree classifier methodology," IEEE Trans. Syst. Man Cybernet., vol. 21, pp. 660–674, 1990.
114. D.A. Freedman, Statistical Models: Theory and Practice; Cambridge University Press: New York, NY, USA, 2009.
115. M.H. Alizai, O. Landsiedel, J.Á.B. Link, S. Götz, K. Wehrle, "Bursty traffic over bursty links," in Proceedings of the 7th ACM Conf on Embedded Networked Sensor Systems, New York, NY, USA, 14–17 March 2009, pp. 71–84.
116. R. Fonseca, O. Gnawali, K. Jamieson, P. Levis, "Four-BitWireless link estimation," in Proceedings of the Sixth Workshop on Hot Topics in Networks (HotNets), Atlanta, GA, USA, 14–15 November 2007.
117. V.N. Vapnik, Statistical Learning Theory; Wiley: New York, NY, USA, 1998

118. M.M. Ramon, T. Atwood, S. Barbin, C.G. Christodoulou, "Signal classification with an SVM-FFT approach for feature extraction in cognitive radio," in Proceedings of the 2009 SBMO/IEEE MTT-S International Microwave and Optoelectronics Conference (IMOC), Belem, Brazil, 3–6 November 2009, pp. 286–289.

119. D.T. Larose, k-Nearest Neighbor Algorithm. In Discovering Knowledge in Data: An Introduction to Data Mining; John Wiley & Sons, Inc.: New York, NJ, USA, 2005; pp. 90–106.

120. D. Son, B. Krishnamachari, J. Heidemann, Experimental study of concurrent transmission in wireless sensor networks. In Proceedings of the 4th International Conference on Embedded Networked Sensor Systems; ACM: New York, NY, USA, 2006; pp. 237–250.

121. G. Trigeorgis, F. Ringeval, R. Brueckner, E. Marchi, M.A. Nicolaou, B. Schuller, and S. Zafeiriou, "Adieu features? End-to-end speech emotion recognition using a deep convolutional recurrent network," in 2016 IEEE international conference on acoustics, speech, and signal processing (ICASSP). IEEE, 2016, pp. 5200–5204.

122. Y. LeCun, Y. Bengio, and G. Hinton, "Deep learning," Nature, vol. 521, no. 7553, pp. 436, 2015.

123. V. Sze, Y.-H. Chen, T.-J. Yang, and J.S. Emer, "Efficient processing of deep neural networks: A tutorial and survey," Proceedings of the IEEE, vol. 105, no. 12, pp. 2295–2329, 2017.

124. R. Hecht-Nielsen, Theory of the backpropagation neural network, in Neural Networks Percep. Elsevier, 1992, pp. 65–93.

125. F. Agostinelli, M. Hoffman, P. Sadowski, and P. Baldi, "Learning activation functions to improve deep neural networks," arXiv preprint arXiv:1412.6830, 2014.

126. J. Schmidhuber, "Deep learning in neural networks: An overview," Neural Networks, vol. 61, pp. 85–117, 2015.

127. H. Xiao, K. Rasul, and R. Vollgraf, "Fashion-MNIST: A novel image dataset for benchmarking machine learning algorithms," arXiv preprint arXiv:1708.07747, 2017.

128. X.J. Zhu, "Semi-supervised learning literature survey," University of Wisconsin-Madison Department of Computer Sciences, Tech. Rep., 2005.

129. A. Radford, L. Metz, and S. Chintala, "Unsupervised representation learning with deep convolutional generative adversarial networks," arXiv preprint arXiv:1511.06434, 2015.

130. V. Mnih, K. Kavukcuoglu, D. Silver, A. Graves, I. Antonoglou, D. Wierstra, and M. Riedmiller, "Playing Atari with deep reinforcement learning," arXiv preprint arXiv:1312.5602, 2013.

131. H. Bourlard and Y. Kamp, "Auto-association by multilayer perceptrons and singular value decomposition," Biol. Cybernetics, vol. 59, no. 4-5, pp. 291–294, 1988.

132. A. Krizhevsky, I. Sutskever, and G. E. Hinton, "Imagenet classification with deep convolutional neural networks," in Advances in neural information processing systems, 2012, pp. 1097–1105.

133. T. Mikolov, M. Karafiát, L. Burget, J. Černocký, and S. Khudanpur, "Recurrent neural network-based language model," in Eleventh annual conference of the international speech communication association, 2010.

134. H.B. McMahan, E. Moore, D. Ramage, and B. A. y Arcas, "Federated learning of deep networks using model averaging," 2016.

135. C.J. Burges, "A tutorial on support vector machines for pattern recognition," Data Mining Knowl. Disc., vol. 2, no. 2, pp. 121–167, 1998.

136. R.H. Myers and R.H. Myers, Classical and modern regression with applications. Duxbury Press Belmont, CA, 1990, vol. 2.

137. Y. Liu, Y. Kang, X. Zhang, L. Li, Y. Cheng, T. Chen, M. Hong, and Q. Yang, "A communication efficient vertical federated learning framework," arXiv preprint arXiv:1912.11187, 2019.

138. T. Lin, S.U. Stich, K.K. Patel, and M. Jaggi, "Don't use large minibatches, use local sgd," arXiv preprint arXiv:1808.07217, 2018.
139. "Tensorflow Federated," TensorFlow, 2022. [Online]. Available: https://www.tensorflow.org/federated. [Accessed: 11-Nov-2022].
140. T. Ryffel, A. Trask, M. Dahl, B. Wagner, J. Mancuso, D. Rueckert, and J. Passerat-Palmbach, "A generic framework for privacy preserving deep learning," arXiv preprint arXiv:1811.04017, 2018.
141. A. Go, R. Bhayani, and L. Huang, "Sentiment140," Site Functionality, 2013c. [Online]. Available at: http://help.sentiment140.com/site-functionality. Abruf am, vol. 20, 2016.
142. S. Caldas, P. Wu, T. Li, J. Konečný, H. B. McMahan, V. Smith, and A. Talwalkar, "Leaf: A benchmark for federated settings," arXiv preprint arXiv:1812.01097, 2018.
143. X. Wang, Y. Han, C. Wangm, Q. Zhao, X. Chen, and M. Chen, "In-Edge AI: Intelligentizing mobile edge computing, caching and communication by federated learning," IEEE Netw., vol. 33, no. 5, pp. 156–165, Sep./Oct. 2019.
144. B. Liu, L. Wang, M. Liu, and C.-Z. Xu, "Federated imitation learning: A novel framework for cloud robotic systems with heterogeneous sensor data," IEEE Robot. Autom. Lett., vol. 5, no. 2, pp. 3509–3516, Apr. 2020.
145. P. Zhou, K. Wang, L. Guo, S. Gong, and B. Zheng, "A privacy preserving distributed contextual federated online learning framework with big data support in social recommender systems," IEEE Trans. Knowl. Data Eng., early access, Aug. 20, 2019, doi: 10.1109/TKDE.2019.2936565.
146. B. Yin, H. Yin, Y. Wum, and Z. Jiang, "FDC: A secure federated deep learning mechanism for data collaborations in the Internet of Things," IEEE Internet of Things J., early access, Jan. 15, 2020, doi: 10.1109/JIOT.2020.2966778.
147. Y. Lu, X. Huang, Y. Dai, S. Maharjan, and Y. Zhang, "Blockchain and federated learning for privacy-preserved data sharing in industrial IoT," IEEE Trans. Ind. Informat., vol. 16, no. 6, pp. 4177–4186, Jun. 2020.
148. J. Ren, H. Wang, T. Hou, S. Zheng, and C. Tang, "Federated learning-based computation offloading optimization in edge computing supported Internet of Things," IEEE Access, vol. 7, pp. 69194–69201, Jun. 2019.
149. R. Fantacci and B. Picano, "Federated learning framework for mobile edge computing networks," CAAI Trans. Intell. Technol., vol. 5, no. 1, pp. 15–21, Mar. 2020.
150. X. Lu, Y. Liao, P. Lio, and P. Hui, "Privacy-preserving asynchronous federated learning mechanism for edge network computing," IEEE Access, vol. 8, pp. 48970–48981, Mar. 2020.
151. M. Yan, B. Chen, G. Feng, and S. Qin, "Federated cooperation and augmentation for power allocation in decentralized wireless networks," IEEE Access, vol. 8, pp. 48088–48100, Mar. 2020
152. N. I. Mowla, N. H. Tran, I. Doh, and K. Chae, "Federated learning based cognitive detection of jamming attack in flying Ad-Hoc network," IEEE Access, vol. 8, pp. 4338–4350, Dec. 2019.
153. T.D. Nguyen, S. Marchal, M. Miettinen, H. Fereidooni, N. Asokan, and A.-R. Sadeghi, "DÏoT: A federated self-learning anomaly detection system for IoT," in Proc. IEEE Int. Conf. Distrib. Comput. Syst., 2019, pp. 756–767.
154. Y.M. Saputra, D.T. Hoang, D.N. Nguyen, E. Dutkiewicz, M.D. Mueck, and S. Srikanteswara, "Energy demand prediction with federated learning for electric vehicle networks," in Proc. IEEE Global Commun. Conf., 2019, pp. 1–6.
155. D. Verma, G. White, and G. de Mel, "Federated AI for the enterprise: A web services-based implementation," in Proc. IEEE Int. Conf. Web Services, 2019, pp. 20–27.
156. Z. Yu et al., "Federated Learning Based Proactive Content Caching in Edge Computing," 2018 IEEE Global Communications Conference (GLOBECOM), 2018, pp. 1–6, doi: 10.1109/GLOCOM.2018.8647616.

157. K. Sozinov, V. Vlassov, and S. Girdzijauskas, "Human activity recognition using federated learning," in Proc. IEEE Int. Conf Parallel Distrib. Process. Appl., Ubiquitous Comput. Commun., Big Data Cloud Comput., Soc. Comput. Netw., Sustain. Comput. Commun., 2019, pp. 1103–1111.

158. W. Zhou, Y. Li, S. Chen, and B. Ding, "Real-time data processing architecture for multi-robots based on differential federated learning," in Proc. IEEE SmartWorld, 2018, pp. 462–471.

159. R. Doku, D.B. Rawat, and C. Liu, "Towards federated learning approach to determine data relevance in big data," in Proc. IEEE Int. Conf. Inf. Reuse Integration, 2019, pp. 184–192.

160. Y. Zhang, Y. Lu, X. Huang, K. Zhang, and S. Maharjan, "Blockchain empowered asynchronous federated learning for secure data sharing in internet of vehicles," IEEE Trans. Veh. Technol., vol. 69, no. 4, pp. 4298–4311, Apr. 2020, doi: 10.1109/TVT.2020.2973651.

161. S. Li, Y. Cheng, Y. Liu, W. Wang, and T. Chen, "Abnormal Client Behavior Detection in Federated Learning," arXiv:1910.09933, Dec. 2019.

162. D. Ye, R. Yu, M. Pan, and Z. Han, "Federated learning in vehicular edge computing: A selective model aggregation approach," IEEE Access, vol. 8, pp. 23920–23935, Jan. 2020.

163. Y. Lu, X. Huang, Y. Dai, S. Maharjan, and Y. Zhang, "Differentially private asynchronous federated learning for mobile edge computing in urban informatics," IEEE Trans. Ind. Informat., vol. 16, no. 3, pp. 2134–2143, Mar. 2020.

164. H. Shiri, J.-H. Park and M. Bennis," Communication-efficient massive UAV online path control: Federated learning meets mean-field game theory," https://arxiv.org/abs/2003.04451v1, Mar. 2020.

165. G. Xu, H. Li, S. Liu, K. Yang, and X. Lin, "VerifyNet: Secure and verifiable federated learning," IEEE Trans. Inf. Forensics Secur., vol. 15, pp. 911–926, Jul. 2019

166. M. Nasr, R. Shokri, and A. Houmansadr, "Comprehensive privacy analysis of deep learning: Passive and active white-box inference attacks against centralized and federated learning," in Proc. IEEE Symp. Secur. Privacy, 2019, pp. 739–753.

167. Z. Wang, M. Song, Z. Zhang, Y. Song, Q. Wang, and H. Qi, "Beyond inferring class representatives: User-level privacy leakage from federated learning," in Proc. IEEE Conf. Comput. Commun., 2019, pp. 2512–2520.

168. A. Triastcyn and B. Faltings, "Federated learning with Bayesian differential privacy," in Proc. IEEE Big Data, 2019, pp. 2587–2596.

169. J. Zhang, B. Chen, S. Yu, and H. Deng, "PEFL: A privacy-enhanced federated learning scheme for big data analytics," in Proc. IEEE Global Commun. Conf., 2019, pp. 1–6.

170. S. Sharma, C. Xing, Y. Liu, and Y. Kang, "Secure and efficient federated transfer learning," in Proc. IEEE Big Data, 2019, pp. 2569–2576.

171. D. Cao, S. Chang, Z. Lin, G. Liu, and D. Sun, "Understanding distributed poisoning attack in federated learning," in Proc. IEEE Int. Conf. Parallel Distrib. Syst., 2019, pp. 233–239.

172. D. Gao, Y. Liu, A. Huang, C. Ju, H. Yu, and Q. Yang, "Privacy preserving heterogeneous federated transfer learning," in Proc. IEEE Big Data, 2019, pp. 2552–2559.

173. K. Bonawitz, V. Ivanov, B. Kreuter, A. Marcedone, H.B. McMahan, S. Patel, D. Ramage, A. Segal, K. Seth, "Practical secure aggregation for federated learning on user-held data," 2016, arXiv:1611.04482. [Online]. Available: https://arxiv.org/abs/1611.04482

174. Y. Zhao, M. Li, L. Lai, N. Suda, D. Civin, V. Chandra, "Federated Learning with Non-IID Data." arXiv 2018, arXiv:1806.00582.

175. K. Kopparapu, E. Lin, "FedFMC: Sequential Efficient Federated Learning on Non-iid Data." arXiv 2020, arXiv:2006.10937.

176. K. He, X. Zhang, S. Ren, and J. Sun, "Deep residual learning for image recognition," in Proceedings of the IEEE conference on computer vision and pattern recognition, 2016, pp. 770–778.
177. L. Wang, W. Wang, and B. Li, "CMFL: Mitigating communication overhead for federated learning."
178. F. Lau, S. H. Rubin, M.H. Smith, and L. Trajkovic, "Distributed denial of service attacks," in proceedings of 2000 IEEE international conference on systems, man, and cybernetics, 2000, pp. 2275–2280.
179. W. Xu, K. Ma, W. Trappe, and Y. Zhang, "Jamming sensor networks: Attack and defense strategies," IEEE network, vol. 20, no. 3, pp. 41–47, 2006.
180. M. Strasser, C. Popper, S. Capkun, and M. Cagalj, "Jamming-resistant key establishment using uncoordinated frequency hopping," in 2008 IEEE Symposium on Security and Privacy (sp 2008). IEEE, 2008, pp. 64–78.
181. Y. Wang, Y. Yang, and T. Luo, "Federated convolutional auto-encoder for optimal deployment of UAVs with visible light communications," in 2020 IEEE International Conference on Communications Workshops (ICC Workshops), Dublin, Ireland, Jun. 2020, pp. 1–6.
182. W.Y.B. Lim, J. Huang, Z. Xiong, J. Kang, D. Niyato, X.S. Hua, C. Leung, C. Miao, "Towards federated learning in UAV-enabled internet of vehicles: A multi-dimensional contract-matching approach," IEEE Trans. Intell. Transp. Syst., vol. 22, pp. 5140–5154, 2021.

4 Application of Wireless Network Data Driven using Edge Computing and Deep Learning in Intelligent Transportation

Zhihan Lv
Faculty of Arts, Uppsala University, Sweden

CONTENTS

DOI: 10.1201/9781003216971-5

4.1 INTRODUCTION

Network data are growing explosively due to intelligent applications in various fields. For example, increasingly diversified services such as the Internet of Vehicles, machine to machine, intelligent grids, and smart cities are growing on a large scale, requiring a deep integration with mobile wireless communication. As a result, people have increasing demands for wireless spectrum resources, leading to wireless spectrum congestion. A proliferation of advanced technologies has been introduced into the gradually popularized fifth-generation (5G) network to improve the performance of communication networks, such as the massive multiple-input-multiple-output technology, multi-access technology, and network slicing [1–3]. In addition, wireless communication networks face a sharp increase in traffic and connection demand. Therefore, many scholars in related fields have focused on combining wireless communication networks and data-driven intelligence.

At present, wireless networks have realized intellectual development in various industries. In intelligent transportation, wireless communication technology can implement the collection of vehicle state information, the perception of driving environment information, and the planning and tracking of vehicle trajectory. It also plays a critical role in the monitoring and regulating traffic flow, the coordinated management and control of vehicles, and the detection and release of information such as pavement and road abnormalities. In a smart city, wireless networks can collect information about urban residents' needs to raise their quality of life [4]. In addition, wireless networks are used in more applications, such as smart homes, smart grids, intelligent waste recycling management, and intelligent lighting. All these applications are combined to develop smart cities together. All Internet of Things (IoT) applications shall be applicable to the heterogeneous network (HetNet) [5]. Applying wireless network technology to medical sensor systems of smart medical treatment can intelligently collect, transmit, and browse data and guarantee data security. Therefore, data-driven intelligence in wireless networks is of great significance.

Of course, artificial intelligence (AI) technologies, such as deep learning (DL) and edge computing (EC), are essential for realizing intelligent data-driven wireless networks. As a computing center among cloud servers and edge devices, EC has the functions of computing, application, storage, and communication. The cloud data center of mobile edge computing and caching (MECC) technology is deployed on the access network side near the data source in a distributed manner. It aims to integrate Internet Service Providers, mobile operators, and IoT devices and perform multiple operations near the data source, such as service awareness, data transmission, information processing, and control optimization. In this way, it shortens the end-to-end transmission delay of data and alleviates the data-processing pressure of the cloud computing platform [6, 7]. Therefore, EC has become one of the critical technologies to interconnect all things and satisfy delay-sensitive applications. Furthermore, the DL algorithm can automatically learn multi-level features from the originally collected data. It can also complete the classification task according to the known essential elements to achieve data-driven intelligence in wireless networks and ensure the effective operation of intelligent network systems [8].

In conclusion, intelligent data-driven management in wireless networks is critical and practical for coping with the diverse IoT application requirements, the explosive growth of mobile data traffic and network-attached devices, and data transmission delay. The research innovation lies in the intelligent prediction and analysis of the massive end-user data collected from the HetNet of the IoT system in intelligent transportation. EC is introduced into the deep convolution random forest neural network (DCRFNN) so that the mobile terminal can directly interact with the edge server through wireless communication. This scheme achieves the cooperative transmission of data and evenly schedules resource loads to provide a basis for subsequent wireless network decision-making, management, and data transmission security.

4.2 RECENT RELATED WORK

4.2.1 STATUS OF THE INTELLIGENT DEVELOPMENT OF WIRELESS NETWORKS

Fifth-generation (5G) communication has become a universal wireless network with its intelligent and adaptive sensing ability, which is significant for practical data-driven intelligence in many fields, such as intelligent transportation and medical treatment. Many scholars have researched smart wireless networks. Zhang et al. proposed a cognitive Internet of Things (CIoT) as the new network paradigm, including big data perception and efficient computing and storage of CIoT edges. Besides, the authors integrated DL with big data analytics to improve the operation efficiency of the system [9]. Sodhro et al. developed an adaptive green (i.e., energy-saving) algorithm based on 5G. In addition, under a specific active time slot, all data packets must be transmitted with the minimum high reliability transmission power. Under this constraint, the authors optimized energy and reliability from the perspective of received signal strength and packet loss rate. They experimented and proved that this cross-layer method achieved a green and reliable platform [10]. Lv et al. analyzed the dual-channel architecture defined by wireless sensor software in 6G/Internet of Everything (IoE) to apply 6G/IoE to various real-life scenes. Moreover, they put forward a reasonable solution to reducing signal interference to transmit relevant signals. The simulation results indicated that the dual-channel structure of the wireless sensor defined by software could transmit the control messages and nodes' sensor messages, respectively. Besides, this structure reduced the traffic load of single channels and avoided the conflict between control messages and sensor messages [11]. Aiming at the potential of adaptive and self-fine tuning in modern communication systems, Ateeq et al. collected an extensive range of data sets, covering various settings of IoT driven by wireless sensor networks. Finally, they determined the correlation between communication parameters and quality of service (QoS) measurement through statistical analysis, laying a foundation for designing a data-driven framework for predictive QoS control in IoT [12].

4.2.2 APPLICATION STATUS OF THE AI ALGORITHM IN DATA-DRIVEN INTELLIGENCE

At present, smart city data are expanding with the rapid development of the Internet. Many scholars have studied the application of AI algorithms such as DL and EC to realize intelligent data-driven wireless networks. Chen et al. proposed an IoT-based manufacturing architecture via EC technology. They analyzed the role of EC from four aspects: edge devices, network communication, information fusion, and cooperation mechanism with cloud computing, and finally provided a technical reference for the deployment of EC in intelligent factories [13]. Li et al. built a hybrid computing framework and designed an intelligent resource scheduling strategy to meet the real-time requirements of intelligent manufacturing supported by EC. They proved that EC achieved good real-time performance, satisfaction, and energy consumption performance in intelligent manufacturing [14]. Yang et al. designed an experimental data-centric traffic flow prediction platform and fused traffic light and vehicle speed models. Finally, through simulation, they validated that the vehicle speed prediction model had the same prediction accuracy as the known long short-term memory model, significantly reduced the computational complexity, and could more effectively capture the real-time changes of traffic conditions [15]. Jaleel et al. proposed a collaborative and adaptive signaling on the edge as a new multi-agent reinforcement learning method to control the phase and timing of traffic signals. They transplanted the controller to the most advanced edge learning platform for performance comparison. The results showed that transplanting the controller to the general platform based on GPU could obtain better real-time performance and decrease the calculation time by eight times [16]. Arthurs et al. analyzed the relationship between cloud computing and intelligent transportation. They believed that cloud computing near the network edge of vehicles and sensors could provide solutions for delay and bandwidth constraints, but the high mobility of vehicles and the heterogeneity of infrastructure remained to be solved [17]. Shahbazi et al. investigated the integration of blockchain, EC, and machine learning and provided advanced data analytics for manufacturing data sets based on swarm intelligence. The experimental results revealed that the EC system significantly reduced the processing time of many tasks in the manufacturing system [18].

According to the research of the scholars mentioned here, the intelligent transmission of collected data has been dramatically improved under the intelligent development trend of wireless networks. Besides, the intelligent processing performance of data has been significantly improved after the intelligent algorithm is introduced into the wireless network. Although the intelligent data processing in wireless networks is analyzed, there is a lack of a comprehensive discussion on the rational utilization of wireless network resources and intelligent data security. Therefore, AI algorithms such as DL and EC are employed for the adaptive scheduling and security performance of wireless network resources in intelligent transportation, which is of great practical significance for wireless network data security and intelligent driving in intelligent transportation.

4.3   APPLICATION ANALYSIS OF THE INTELLIGENT MODEL DRIVEN BY WIRELESS NETWORK DATA BASED ON EC AND DL IN INTELLIGENT TRANSPORTATION

4.3.1   DEMAND ANALYSIS OF DATA-DRIVEN INTELLIGENCE IN WIRELESS NETWORKS OF INTELLIGENT TRANSPORTATION

Today, wireless networks have a widespread application range, such as vehicle real-time location monitoring in the transportation field, real-time sharing of network courses, and real-time broadcasting of sports events of the Olympic Games. However, various IoT applications' data rates and delays are not necessarily the same. Therefore, mobile operators usually improve capacity, coverage, and quality of services to cope with the diversified requirements of IoT applications and the explosive growth of mobile data and network connection devices. Consequently, ultra-dense HetNets come into being. For example, micro base stations (MBSs) are usually deployed for ultra-dense HetNets and installed by users in indoor environments to provide broad mobile coverage and capacity. MBSs are used in densely populated hot spots and public vehicles, such as buses, subways, and trains, to reduce the traffic load of macro cellular networks and improve mobile users' voice and video service experience. Vast quantities of traffic can be unloaded from MBSs to reduce their deployment cost [19, 20]. Deploying MBSs and data-driven intelligence in wireless networks enables multitudes of users to use the wireless network in a given area of the same radio spectrum and improves the regional spectrum efficiency. Figure 4.1 illustrates the layered architecture of the HetNet system with data on various fields.

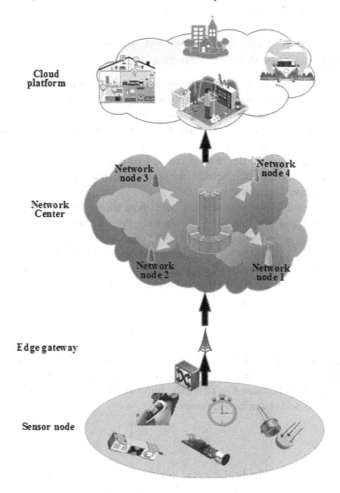

FIGURE 4.1   The layered architecture of the HetNet system with data from various fields.

Through Figure 4.1, there is a wealth of data in IoT HetNet systems, such as intelligent medical treatment, intelligent transportation, smart cities, and smart homes. The operation of an IoT system is inseparable from the intelligent drive of the collected data. The signal data collected from various sources, such as sound, light, mechanics, chemistry, thermal energy, and electric energy, is transmitted to the cloud platform through the network layer. Then, the cloud platform implements complex decision-making algorithms to summarize and analyze the data and executes relevant commands issued by users.

Therefore, in the HetNet system, intelligent drive and security of the collected information have become necessary. With the gradual growth of vehicle users, the amount of data in the intelligent transportation IoT system also increases massively. Therefore, this chapter takes the intelligent IoT system in road network transportation as the research subject to realize the IoT system's effective operation and performance safety of intelligent transportation through data-driven intelligence in wireless networks.

4.3.2 APPLICATION ANALYSIS OF INTELLIGENT ALGORITHMS IN DATA-DRIVEN INTELLIGENCE IN WIRELESS NETWORKS OF INTELLIGENT TRANSPORTATION

In the wireless network of intelligent transportation systems using IoT, the amount of data in HetNets increases sharply with the number of end-users. The communication network center of road network traffic can complete cooperative data transmission in the transportation system by collecting the context data uploaded by vehicles. The context data includes the location of surrounding vehicles, the quality of channels with surrounding vehicles, the data perceived or cached by vehicles, and the perceived data requested by each vehicle. Figure 4.2 displays the ultra-dense wireless network in intelligent transportation.

In the ultra-dense HetNet network of intelligent transportation, the set of context information uploaded by the node N_0 containing the surrounding vehicle is denoted as $\{N_1, N_2, N_3\}$, the signal-to-noise ratio (SNR) of channels connected to surrounding vehicles is $\{SINR_{01}, SINR_{02}, SINR_{03}\}$. Besides, $\{d_1, d_2, d_3\}$ represents the cached data segment set, and $\varnothing$ denotes the data segments that are not cached but need to be downloaded. Based on the SNR between node N_i and node N_j, the channel capacity between them is calculated according to:

$$C_{i,j} = W \log\left(1 + SINR_{i,j}\right) \qquad (4.1)$$

Since the environmental information perceived by the vehicle is in the form of pictures or point clouds, the data size is large. Therefore, the data are first divided into data blocks $\{d_1, d_2, \cdots\}$ of size C_0. Then, the number of data blocks transmitted by the channel between nodes can be expressed as:

$$M_{i,j} = \left\lfloor \frac{C_{i,j} T_0}{C_0} \right\rfloor \qquad (4.2)$$

where T_0 refers to the scheduling cycle, and $\lfloor b \rfloor$ indicates that b is rounded down. The EC center in the HetNet collects the context data of all vehicles. Figure 4.3 reveals the intelligent data transmission link in the edge distributed devices.

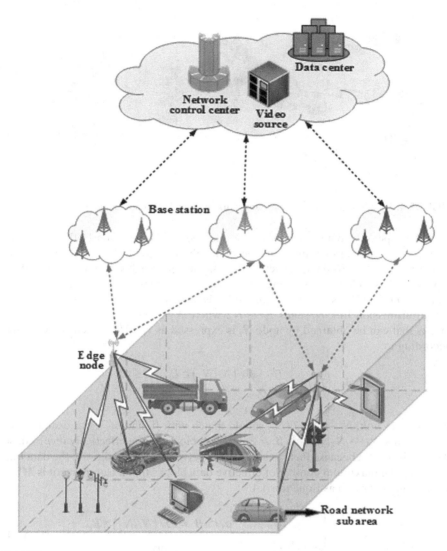

FIGURE 4.2   Data transmission in the ultra-dense HetNet of intelligent transportation.

In Figure 4.3, mobile edge computing (MEC) reduces latency by deploying computing and storage capabilities to the radio access network (RAN) edge, ensuring
effective network operation and reliable service delivery. Consequently, it has sufficient capacity to perform intensive computing and delay critical tasks on mobile
devices. MEC provides cloud computing functions in the RAN, allowing direct
mobile communication between core networks and end-users instead of directly
connecting users to the nearest edge network supporting cloud services. At the
same time, MEC can enhance computing capacity and avoid system bottlenecks and
failures [21].

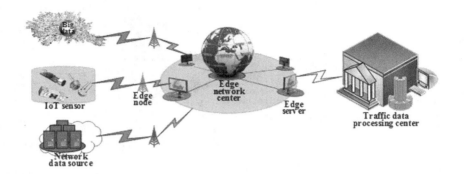

FIGURE 4.3 Intelligent data transmission link in edge distributed devices.

The cooperative data transmission in intelligent transportation HetNets is modeled as the weighted maximum independent subset in graph theory. For each node N_i ($N_i \in N$) that sends data, the set of all possible data block sets sent by the node is represented as $T(N_i)$. The sending data block set that may be sent is denoted as T_k^l, and $T_k^l \in T(N_i)$. The set of receiving nodes is expressed as $0 \leq k < |T(N_i)|$, and the data transmission link N_i sends T_k^l to N_j, where $N_j \in R(T_k^l)$. The set of valid data blocks that can be obtained by node N_j is expressed as $V_{i,j}(T_k^l)$, which is calculated according to:

$$V_{i,j}\left(T_k^l\right) = \begin{cases} T_k^l \cap \beta_j & \left(N_i, N_j\right) \in E_u, \left|T_k^l\right| \leq M_{i,j} \\ \varnothing & Otherwise \end{cases} \quad (4.3)$$

For the data link, $N_i \rightarrow N_j$, $T(N_i)$ should match with β_j and $M_{i,j}$. For example, for the data link $N_4 \rightarrow N_2$, if $\alpha_4 \cap \beta_2 = \varnothing$, no data block can be effectively received, so the link is invalid because there is a mismatch between $T(N_i)$ and β_j. In the data link $N_i \rightarrow N_j$, the maximum number of effective data blocks that it can transmit is $MV_{i,j}$, which can be written as Equation 4.4.

$$MV_{i,j} = \begin{cases} 0 & \alpha_i \cap \beta_j = \varnothing / \left(N_i, N_j\right) \notin E_u \\ \min\left\{|\alpha_i|, M_{i,j}\right\} & Otherwise \end{cases} \quad (4.4)$$

Furthermore, the data link transmission in HetNets is transformed into a directed link matching transmission diagram G_d. Through such conversion, many impractical transmissions in the Intelligence Transportation System (ITS) in the road network area can be removed, which significantly reduces the complexity. In the constructed directed link matching transmission diagram G_d, the nodes represent possible transmission, and the edges between nodes refer to wireless communication restrictions or interference [22]. Then, the neighbor of the node N_i in G_d is indicated as $\Omega(N_i, G_d)$, as shown in Equation 4.5.

$$\Omega\left(N_i, G_d\right) = \left\{N_j | W\, E_{i,j} > 0, N_j \in N\right\} \quad (4.5)$$

In Equation 4.5, $WE_{i,j}$ represents the weight of the edge $N_i \rightarrow N_j$ in the G_d. As $T(N_i)$ is related to $\Omega(N_i, G_d)$, $MV_{i,j}$, and α_j, then T_k^l shall meet Equation 4.6.

$$\begin{cases} T_k^l \subseteq \alpha_i \\ \left|T_k^l\right| \in \left\{MV_{i,j} \mid N_j \in \Omega(N_i, G_d)\right\} \end{cases} \quad (4.6)$$

Then, the undirected interference diagram based on data transmission is constructed, as presented in Equation 4.7.

$$G_c = (U_c, E_c) \quad (4.7)$$

The collaborative data transmission problem is modeled as the maximum independent subset problem. U_c denotes the set of nodes and E_c refers to the set of edges. Each node u_m has a weight $w(u_m) = V(u_m)$, where $u_m = N_i \xrightarrow{T_{k*}^l} \{\alpha\}$. The independent subset in G_c is a set of nodes and any two are not adjacent. In other words, the set $X \subseteq U_c$ is an independent subset. If, and only if, any two nodes u_i and u_j belong to X, Equation 4.8 is satisfied.

$$(u_i, u_j) \notin E_c \quad (4.8)$$

Then, the weight of the independent subset X is expressed as $W(X)$. The maximum independent subset problem is to find X^* ($X^* \subseteq U_c$), and Equation 4.9 shall be met.

$$W(X^*) \geq W(X), \forall X \subseteq U_c \quad (4.9)$$

In other words, the maximum independent subset problem can be modeled as an integer programming problem, as shown in Equation 4.10.

$$\max \sum_{u_k \in X} w(u_k)$$
$$s.t. \begin{cases} X \subseteq U_c \\ (u_n, u_m) \notin E_c, \forall u_n, \forall u_m \in X, u_n \neq u_m \end{cases} \quad (4.10)$$

The balanced scheduling of communication resources is also crucial for the HetNet of ITS. Denote r_i as the grid (m, n) of the road network area and the neighbor of r_i can be expressed as Equation 4.10.

$$R(r_i) = \left\{(x_1, x_2) \mid \max\left\{|m - x_1|, |n - x_2|\right\} = 1\right\} \quad (4.11)$$

In Equation 4.10, $0 \leq x_1 < M$ and $0 \leq x_2 < N$. For each grid r_i, assume that there is an MBS denoted as $B(r_i)$, used as the deployment node of EC. Since different base stations can serve vehicles in different amounts, some base stations may be congested, and others are idle. At time t, the traffic flows from r_i to r_j ($r_j \in R(r_i)$) and the traffic flow of r_j can be expressed as:

$$S_{t,r_i \rightarrow r_j} = f_{in,t}(r_i) \cap S_t(r_i) \cap S_t(r_j) + S_t(r_i) \cap S_{t+1}(r_j) \quad (4.12)$$

The number of vehicles and the average speed of the traffic flow $S_{t,r_i \to r_j}$ are calculated according to:

$$N_{t,r_i \to r_j} = \left| S_{t,r_i \to r_j} \right| \tag{4.13}$$

$$V_{t,r_i \to r_j} = \sum_{\alpha \in S_{t,r_i \to r_j}} \frac{v_t(\alpha)}{N_{t,r_i \to r_j}} \tag{4.14}$$

where $v_t(\alpha)$ refers to the speed of the vehicle α at time t. When the vehicle α passes through the network area grid r_i, the total amount of data that the EC center needs to transmit to the vehicle at any time is $CV_t(r_i)$, which can be presented as Equation 4.15.

$$CV_t(r_i) = \sum_{r_j \in R(r_i)} \sum_{\alpha \in S_{t,r_i \to r_j}} \beta_\alpha(r_i) \tag{4.15}$$

For the MBS $B(r_i)$, the total amount of data that can be downloaded within its coverage is expressed as $CS_t(r_i)$, related to the base station's coverage, the distribution of vehicles, and the number and distribution of small cells or MBSs. Due to the mobility and uneven distribution of vehicles, a base station may not meet all data download requests, that is, $CV_t(r_i) > CS_t(r_i)$. At this time, the base station $B(r_j)$ $(r_j \in R(r_i))$ near $B(r_i)$ can unload some data to balance the load of the base station.

When unloading the load of the base station, the vehicle $S_{t,r_i \to r_j}$ from r_j to r_i is allowed to download the high-precision map corresponding to grim r_i. The amount of data to be unloaded can be expressed as $T_{t,r_i \to r_j}$. At this time, the total amount of data to be downloaded by the base station $B(r_i)$ is:

$$h_i\big(B(r_i)\big) = CV_t(r_i) + \sum_{r_j \in R(r_i)} \big(T_{t,r_j \to r_i} - T_{t,r_i \to r_j} \big) \tag{4.16}$$

Then, the utility function of the MBS can be written as Equation 4.17.

$$h_i^*\big(B(r_i)\big) = \begin{cases} 0 & h_i\big(B(r_i)\big) < CS_t(r_i) \\ h_i\big(B(r_i)\big) - CS_t(r_i) & Otherwise \end{cases} \tag{4.17}$$

Based on the utility function defined here, the load balancing problem in 5G HetNets can be defined as Equation 4.18.

$$\min \sum_{r_i} h_i^*\big(B(r_i)\big) + \sum_{r_i} \sum_{r_j \in R(r_i)} \left| T_{t,r_j \to r_i} \right|$$

$$s.t. \begin{cases} T_{t,r_i \to r_j} \geq 0, r_j \in R(r_i) \\ T_{t,r_i \to r_j} \leq T_{t,r_i \to r_j} \cdot D(r_i, r_j) / V_{t,r_i \to r_j} \cdot C_0 \end{cases} \tag{4.18}$$

In Equation 4.18, C_0 represents the maximum rate of a communication connection between the vehicle and the EC center and $D(r_i, r_j)$ denotes the distance between the centers of r_i and r_j.

4.3.3 INTELLIGENT DATA-DRIVEN MODEL FOR WIRELESS NETWORKS OF INTELLIGENT TRANSPORTATION BASED ON DL AND EC

Because of the massive amount of end-user data collected in the HetNet of the IoT system in intelligent transportation, this chapter aims to achieve the cooperative transmission of data and real-time load balancing scheduling of resources. Meanwhile, the mobile terminal can directly interact with the edge server through wireless communication after combining DCRFNN with EC [23]. Under the constraints of the limited total storage capacity of the edge server, a reasonable data placement strategy is conducive to reducing the transmission time of alarm results to meet the calculation and transmission requirements of ultra-low delay of traffic accident alarms. EC can effectively unload and deploy computing-intensive tasks of traffic accident prediction to edge devices, which simplifies the tasks transmitted to the cloud and releases network bandwidth resources. Figure 4.4 shows the intelligent data-driven model of intelligent transportation IoT based on EC and DL.

In this chapter, the load balancing strategy of 5G HetNets is realized through the cooperative transmission of data and the load balancing of communication resources. The data features of road networks in intelligent transportation extracted by DCRFNN reported here consist of the feature extractor and classification prediction. The feature extractor only includes the input layer, convolution layer, pooling layer, and activation function of CNN while eliminating the fully connected layer and softmax classifier of CNN. The random forest (RF) algorithm with multiple decision trees is responsible for classification and prediction [24]. Firstly, CNN with weight-sharing characteristics processes the massive multi-dimensional traffic data sets without pressure. After the convolution layer extracts the features, the RF algorithm is selected to classify the processed high correlation multi-dimensional features and intelligently drive the data in heterogeneous traffic networks.

Then, the traffic flow feature of DCRFNN is extracted and analyzed. The pooling layer sums the obtained convolution characteristic map in units of eigenvalues in each neighborhood to reduce the overfitting degree of CNN parameters and models. Next, the value is transferred into an eigenvalue, weighted by the scalar $\beta^{(l+1)}$, and added by the offset $b^{(l+1)}$. Finally, a ReLU activation function produces a feature map about $\beta^{(l+1)}$ times smaller. The convolution kernel of the convolution layer is multiplied by the corresponding elements in each input feature map. These products are summed and added to bias. The activation function processes the result as the output. The output $a_{u,v}^{(l)}$ of the convolution layer can be expressed as Equation 4.19.

$$a_{u,v}^{(l)} = f\left(z_{u,v}^{(l)}\right) \tag{4.19}$$

$$z_{u,v}^{(l)} = \sum_{i=-\infty}^{\infty}\sum_{j=-\infty}^{\infty} x_{i+u,j+v}^{(l-1)} G_{roti,v}^{(l)} \chi(i,j) + b^{(l)} \tag{4.20}$$

$z_{u,v}^{(l)}$ in Equation 4.19 can be written as Equation 4.20. The expression of $\chi(i,j)$ is $\chi(i,j) = \begin{cases} 1\,(i \geq 0, j \leq n) \\ 0 \end{cases}$, where $x_{i+u,j+v}^{(l-1)}$ denotes the input characteristic matrix and $f(\cdot)$ refers to the activation function.

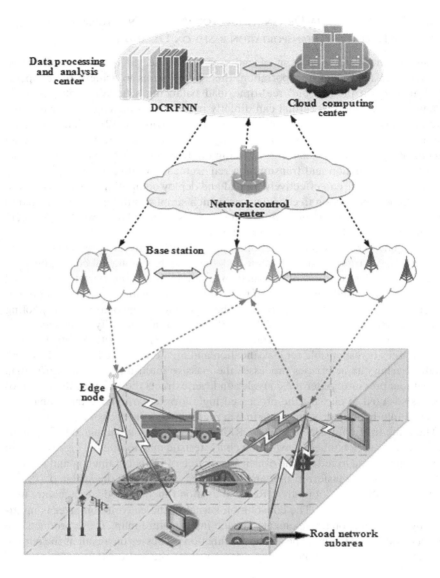

FIGURE 4.4 An intelligent data-driven model of intelligent transportation IoT based on EC and DL.

Moreover, there is an error between the actual output of the network and the theoretical target. Therefore, it is necessary to recurse the error between the actual and target output through backpropagation and calculate the adjustment amount of each weight and offset in turn [25]. The convolution layer's objective function (error-sensitive term) is presented as Equation 4.21.

$$\delta_{g,h}^{(l)} = \frac{\partial J\left(W, b; x, y\right)}{\partial z_{g,h}^{(l)}} = \beta^{(l+1)} \delta_{i+pr, j+qr}^{(l+1)} f'\left(z_{g,h}^{(l)}\right) \tag{4.21}$$

In Equation 4.21, $\beta^{(l+1)}$ refers to the weight of each unit of the convolution kernel. The objective function (error-sensitive term) of the pooling layer (sub-sampling layer) can be expressed as Equation 4.22.

$$\delta_{g,h}^{(l+1)} = \frac{\partial J(W,b;x,y)}{\partial z_{g,h}^{(l+1)}} = \sum_{u=0}^{q-1}\sum_{v=0}^{q-1} \delta_{g,h}^{(l+2)} G_{rotg-u,h-v}^{(l+2)} f'\left(z_{g,h}^{(l+1)}\right) \tag{4.22}$$

The adjustment amount of weight and offset of the convolution layer is:

$$\frac{\partial J(W,b;x,y)}{\partial b^{(l)}} = \sum_{u=0}^{m-n}\sum_{v=0}^{m-n} x_{rotg+u,h+v}^{(l-1)} \delta_{u,v}^{(l)} \tag{4.23}$$

Then, the weight and offset are adjusted in turn according to the adjustment amount to make the actual output tend to a fixed value and constantly approach the actual value. The expression ability of the linear model is poor. Therefore, the activation function ReLU, i.e., $Re\,LU(x) = \begin{cases} x\,(x>0) \\ 0\,(x \le 0) \end{cases}$ is introduced to add nonlinear factors in the convolution process, responsible for obtaining maximal $(0, x)$. When $x \le 0$, the output of the activation function ReLU is 0. A large number of neurons outputting 0 indicates a great sparsity, high representativeness of the extracted features, and a strong generalization ability. It means a small number of neurons in action and calculation, ultimately realizing parameter sparsity.

When the extracted traffic data set contains highly relevant data features that may induce traffic accidents, it is necessary to reshape the extracted features to a size suitable for the input of the RF classifier. First, the bootstrap sampling method selects k self-help sample sets from the input feature map as the training test set to construct an RF classifier. Then, m features are randomly selected from M features, and one feature is selected as the splitting feature of the node until it can no longer be split. Finally, the previous operation is repeated n times to obtain n classification and regression trees (CARTs) to establish the RF classifier [26]. The RF classifier consists of several same but independent CARTs. The CART algorithm arranges the real-time traffic data in ascending order and calculates the heterogeneity of the output variable values in groups. The heterogeneity is the Gini coefficient to measure the error measurement of the tree. The smaller the Gini coefficient, the higher the purity of the data set. The Gini coefficient can be expressed as Equation 4.24.

$$gini(T) = 1 - \sum p_j^2 = 1 - \sum \left(\frac{n_j}{S}\right)^2 \tag{4.24}$$

In Equation 4.24, p_j signifies the frequency of category j in the sample set T, n_j represents the number of category j in the sample set T, and S denotes the number of samples in the sample set T. After introducing variables, the Gini coefficient can be described as Equation 4.25.

$$gini_{sample}(T) = \frac{S_1}{S_1 + S_2} gini(T_1) + \frac{S_2}{S_1 + S_2} gini(T_2) \tag{4.25}$$

In Equation 4.25, S_1 and S_2 refer to the number of samples after being divided into two types of samples, and $gini(T_1)$ and $gini(T_2)$ are the Gini coefficient of two types of samples. Finally, the RF classifier outputs the prediction risk of traffic accidents, i.e., a probability value ranging between 0 and 1. Traffic accidents are severe if the predicted value is more biased to 1. On the contrary, the risk of traffic accidents is low.

Furthermore, in the actual road network planning scenario, due to the social attributes of users, the distribution density of users fluctuates with the changes of time and space, such as commuting during peak hours and non-peak periods. In peak hours, it is feasible to confirm the minimum number of base stations and optimal locations meeting the current user's needs, then deploy them in the corresponding locations and open them. Similarly, in the off-peak period, it is necessary to determine the minimum number of base stations and their optimal location to meet the current user's needs. Some small base stations have the exact location as those in the peak hours, so they only need to be opened without deployment again [27]. For this adjustable scheme, the coverage radius of each edge gateway is set as r. Any edge gateway e can cover multiple terminal traffic generators. Each terminal traffic generator t can also be within the coverage of multiple edge gateways simultaneously, but the nearest edge gateway can only serve each terminal traffic generator t. Here, the binary variable C_{te} is selected to indicate whether the edge gateway e covers the terminal traffic generator t, then there is:

$$C_{te} = \begin{cases} 1 & if d_{te} \leq r \\ 0 & if d_{te} > r \end{cases} \quad (4.26)$$

The binary variable I_{te} refers to whether the edge gateway e serves the terminal traffic generator t. Therefore, there is:

$$I_{te} = \begin{cases} 1 & if d_{te} \leq \min\{d_{t1}, d_{t2}, \cdots, d_{tn}\} \\ 0 & else \end{cases} \quad (4.27)$$

Where n refers to the number of edge gateways that cover the terminal traffic generator in the road network. Therefore, each terminal traffic generator must meet the coverage condition shown in Equation 4.28 to ensure the reliability of coverage.

$$\sum_{\forall e \in E} C_{te} \geq 1, \forall t \in T \quad (4.28)$$

In addition, the premise that the edge gateway e serves the terminal flow generator t in the road network is that the edge gateway e covers the terminal flow generator t, so the edge gateway e needs to be deployed in advance. Therefore, the binary variables I_{te}, C_{te}, and ε_e should satisfy Equation 4.29.

$$I_{te} \leq C_{te} \leq \varepsilon_e, \forall t \in T, \forall e \in E \quad (4.29)$$

In Equation 4.29, $I_{te}, C_{te}, \varepsilon_e \in \{0,1\}$, $\forall t \in T$, and $\forall e \in E$.

Then, the location of the edge gateway is evaluated by delay. The placement of the edge gateway principally affects the capacity allocation and load balancing. Consequently, it affects the unloading delay of the computing task, including the transmission delay of the task from the intelligent terminal to the edge gateway and the processing delay of the edge gateway [28]. The delay of unloading the calculation task to the edge gateway is calculated according to Equation 4.30.

$$t_{EN} = t_{tra} + t_{exe} \qquad (4.30)$$

In Equation 4.30, t_{EN} refers to the total delay, t_{tra} stands for the transmission delay from the intelligent terminal to the edge gateway, and t_{exe} represents the processing delay of the edge gateway. In addition, the uplink data rate at which the terminal traffic generator unloads the calculation task locally to the edge gateway through the wireless channel is calculated according to Equation 4.31.

$$R_t = W \log_2 \left(1 + \frac{P_t H_{te}}{N_0 + \sum_{q \in T/(t)} P_q H_{qe}} \right) \qquad (4.31)$$

In Equation 4.31, W refers to the channel bandwidth, P_t signifies the transmission power of the terminal traffic generator t. H_{te} represents the channel gain between the terminal traffic generator t and the edge gateway e, and N_0 denotes the background noise power.

Then, the computing task of the terminal traffic generator is modeled as $M_t = (D_t, C_t)$, which can be unloaded to the edge gateway by the terminal traffic generator. In $M_t = (D_t, C_t)$, D_t refers to the size of the input data, and C_t represents the central processing unit (CPU) cycles required to complete the computing task of the terminal traffic generator. For the unloading of computing tasks, the terminal traffic generator transmits the input data used for computing tasks to the edge gateway through wireless access, producing a transmission delay. The transmission delay $t_{te}^{tra} = \frac{D_t}{R_t}$ of the terminal traffic generator t unloading the input data of size D_t to the edge gateway e can be calculated through the uplink data rate in the wireless channel.

After unloading tasks, the edge gateway processes the computation task $M_t = (D_t, C_t)$. Here, the computing power of the edge gateway e, that is, CPU cycles per second, is defined as f_e. Thus, the execution time of the task M_t generated by the terminal traffic generator t on the edge gateway e is $t_{te}^{exe} = \frac{C_t}{f_e}$. Finally, the total delay of unloading the calculation task by the terminal traffic generator t to the edge gateway e is:

$$t_{te} = t_{te}^{tra} + t_{te}^{exe} \qquad (4.32)$$

Algorithm 1 provides the algorithm flow of EC and DCRFNN in the intelligent transportation IoT intelligent data-driven model.

Algorithm 1: Algorithm flow of the combination of EC and DCRFNN

---

1   **Start**

2   Input: the current and historical traffic situation $\{F_h \mid h = t - l, t - l + 1, \cdots, t\}$, the real traffic situation
     in the future $\{F_k \mid k = t + 1, t + 2, \cdots, t + s\}$, the surrounding vehicles gather $\{N_1, N_2, N_3\}$, SNR of
     surrounding vehicle channel $\{SINR_{01}, SINR_{02}, SINR_{03}\}$, collection of cached data fragments
     $\{d_1, d_2, d_3\}$.

3   Output: Future traffic situation F_k^*

4   Initialization of variable parameters

5   **If** $CV_t(r_i) > CS_t(r_i)$ **then**

6   **For** an Acer station $B(r_i)$

7   At time t, the traffic flow from r_i to r_j can be expressed as Equation 4.12

8   The number of vehicles in the traffic flow is expressed by Equation 4.13

9   The average speed of traffic flow is expressed by Equation 4.14

10   The total amount of data that the EC center needs to transmit to the vehicle at time t is expressed by
     Equation 4.15

11   **end** for

12   Unload part of the data, the total amount of data can be expressed by Equation 4.16

13   The utility function of Acer station K is expressed by formula (4.17)

14   Prediction of traffic situation based on Deep convolution random forest neural network

15   **end** if

16   **Until** Communication resource data load balancing

17   **End**

4.3.4 CASE ANALYSIS

The SUMO simulation platform [29] is selected for model construction to verify the data-driven model of the IoT-based intelligent transportation system using EC and DL. The experimental equipment is a computer with an IS-7300HQ processor, 12 GB memory, and an NVIDIA GeForce GTX 1050 graphics card. DCRFNN predicts traffic accident risks on a DL framework named Keras Neural Network Library based on Theano [30]. The simulation experiment uses Python for programming. This chapter takes the standard IoT system in intelligent transportation as a case for analysis. The vehicle terminal data used in the experiment is collected from March through July of 2020. The total number of vehicles is 200 to 700, the step size is 100, and the simulation cycle is 300 seconds. The roadside unit and the vehicle use IEEE 802.11p to communicate with each other, and all vehicles upload task data at a minimum data rate of 3 Mbps to provide the best communication reliability. The communication distance of the vehicle is 200 meters, and the communication distance of the roadside unit is 600 meters. Finally, the obtained data are divided into the training and test data sets by 8:2.

The intelligent data-driven model based on EC and DCRFNN is compared with other advanced DL algorithms proposed by scholars in related fields, including AlexNet [31], DenseNet [32], Visual Geometry Group Network (VGGNet) [33], Interleaved Group Convolutions for Deep Neural Network (IGCNet) [34], and ResNet [35]. The convergence of loss functions and recognition accuracy of these algorithms are analyzed to evaluate the models' performance. Moreover, the hyper-parameters of the neural networks are as follows: the number of iterations is 60, the simulation time is 2,000 seconds, and the batch size is 128. Furthermore, the contrastive analysis focuses on security performance and the time delay of data transmission.

The performance of each algorithm is compared from the perspectives of accuracy (*Acc*), precision (*Pre*), recall (*Rec*), and F1-value. *ACC* measures the overall classification accuracy, that is, the ratio of correctly predicted samples; *Rec* evaluates the coverage of positive samples, that is, the proportion of correctly classified positive samples in the total number of positive samples; *Pre* represents the ratio of the example classified as a positive example to the actual positive example. The most commonly used method is the F1-value, the weighted harmonic average of *Pre* and *Rec*. *Acc*, *Pre*, *Rec*, and F1-value are calculated according to:

$$Acc = \frac{\sum\limits_{i=1}^{l} \frac{TP_i + TN_i}{TP_i + FP_i + TN_i + FN_i}}{l} \tag{4.33}$$

$$\mathrm{Pr}\,ecision = \frac{\sum\limits_{i=1}^{l} \frac{TP_i}{TP_i + FP_i}}{l} \tag{4.34}$$

$$\mathrm{Re}\,call = \frac{\sum\limits_{i=1}^{l} \frac{TP_i}{TP_i + FN_i}}{l} \tag{4.35}$$

$$F1 = \frac{2\,\mathrm{Pr}\,ecision \cdot \mathrm{Re}\,call}{\mathrm{Pr}\,ecision + \mathrm{Re}\,call} \tag{4.36}$$

where *TP* refers to the number of positive samples predicted to be positive; *FP* stands for the number of negative samples predicted to be positive; *FN* stands for the number of positive samples predicted to be negative; *TN* represents the number of negative samples predicted to be negative.

4.4   RESULTS AND DISCUSSION

4.4.1   ANALYSIS PREDICTION OF THE ACCURACY OF DIFFERENT ALGORITHMS IN WIRELESS NETWORKS OF INTELLIGENT TRANSPORTATION

The intelligent data-driven model, AlexNet, DenseNet, VGGNet, IGCNet, and ResNet are compared from accuracy, precision, recall, and F1-value, respectively, to study the prediction performance of the intelligent data-driven model of intelligent transportation IoT. Figure 4.5 shows the results. In addition, the training and test time required by each algorithm are compared. The results are presented in Figure 4.6.

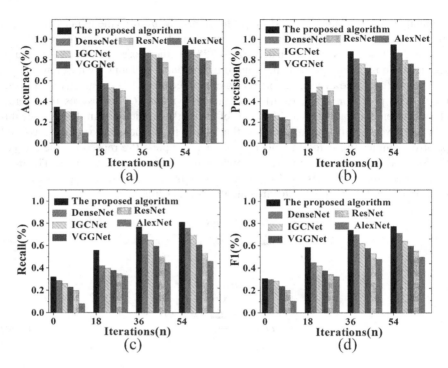

FIGURE 4.5 The influence of iteration times on the prediction accuracy of building vertical images under different algorithms.

(a. Accuracy; b. Precision; c. Recall; d. F1-value)

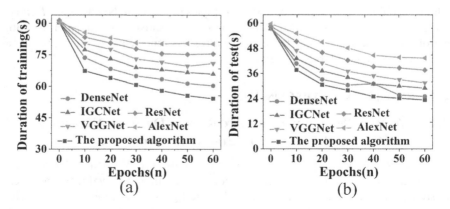

FIGURE 4.6 Comparison of test and training time required by each algorithm.

(a. Training time; b. Test time)

As shown in Figure 4.5, the recognition accuracy of the intelligent data-driven model reaches 94.11%, at least 3.28% higher than that of the algorithms proposed by other scholars. The precision, recall, and F1-value of the intelligent data-driven model are 92.81%, 81.15%, and 77.39%, respectively. Compared with other algorithms, the precision, recall, and F1-value of the model reported here are higher, at least 3.61% higher than other algorithms. The intelligent data-driven model of the intelligent transportation IoT based on EC combined with DL constructed here is superior to algorithms proposed by other scholars in related fields in prediction accuracy.

Figure 4.6 reveals the test and training time required by each algorithm. Through results in Figure 4.6, with the increase in the number of iterations, the required training time and test time show a changing trend of increasing first and then basically unchanged, that is, reaching convergence. Specifically, the intelligent data-driven model's training and test time are stable at 54.18 seconds and 23.39 seconds, respectively, significantly shorter than the other comparative algorithms. Therefore, the intelligent data-driven model for wireless network data in intelligent transportation IoT systems can achieve a higher prediction effect in a shorter time.

4.4.2 ANALYSIS OF SECURITY PERFORMANCE AND TIME DELAY OF DATA TRANSMISSION IN WIRELESS NETWORKS OF INTELLIGENT TRANSPORTATION UNDER DIFFERENT ALGORITHMS

In analyzing the security performance and delay of data transmission, the above comparative algorithms are compared from transmission data volume, end-user number, and throughput. The results are shown in Figures 4.7–4.9.

In Figure 4.7, with the increase in the amount of transmitted data, the average delivery rate of network data presents an upward trend. The delivery rate of data messages of the intelligent data-driven model remains at 77.84% (Figure 4.7-a), and the average network data leakage rate shows no apparent change. Besides, the data message leakage rate of the intelligent data-driven model is the lowest, finally stabilizing at 3.12% (Figure. 4.7-b). In Figure 4.7-c, the ResNet algorithm presents a high packet loss rate, which the problem of the remote terminal may cause. The packet loss rate of the intelligent data-driven model is the lowest, no more than 5.42%, which the equalization of the transmitted data may cause. Therefore, under different amounts of transmission data, the intelligent data-driven model based on EC and DL achieves a high average delivery rate, low average leakage rate, and low packet loss rate, showing excellent data transmission performance in the wireless network.

Figure 4.8 compares the throughput of different algorithms with the growth of end-users (vehicles). The intelligent data-driven model is the highest reception throughput, DenseNet ranks second, and AlexNet is the lowest. Reception throughput refers to the number of packets received from other adjacent end-users. Higher reception throughput means that end-users can obtain more information from other adjacent end-users, meeting the micro perception requirements of end-users. Transmission throughput indicates how many packets can be sent per frame per user. Due to the requirements of related secure applications, the transmission throughput should be about one packet per frame to ensure that each end-user has the opportunity to obtain

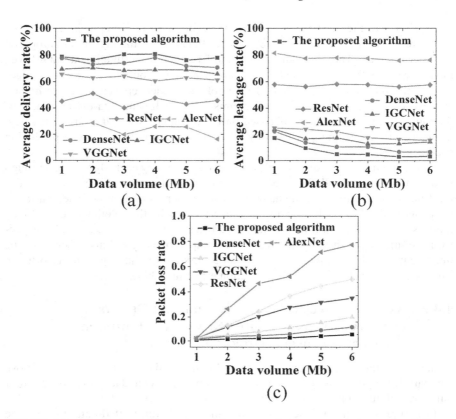

FIGURE 4.7 Comparison of security performance of data transmission under different amounts of transmission data of different algorithms.

(a. Data average delivery rate; b. Data average leakage rate; c. Data packet loss rate)

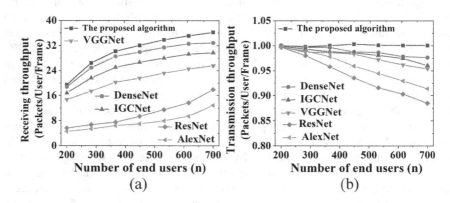

FIGURE 4.8 Curves of throughput with the increase in the number of end-users (vehicles) under different algorithms.

(a. Reception throughput; b. Transmission throughput)

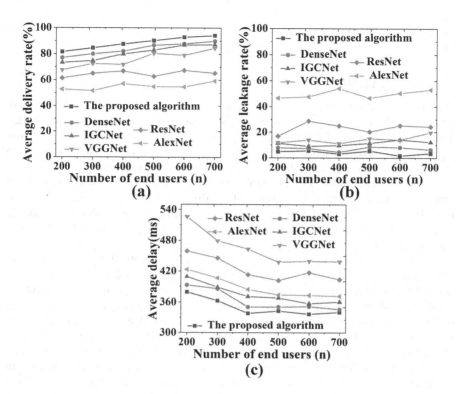

FIGURE 4.9 Comparison of network data security transmission of various algorithms under different numbers of end-users (vehicles).

(a. Average delivery rate; b. Average leakage rate; c. Average delay)

its location. The analysis of transmission throughput suggests that the transmission throughput decreases with the increase in the number of vehicles. It directly leads to some end-users being unable to obtain time slots and cannot meet the stringent requirements of secure applications. Therefore, compared with other protocol algorithms, the intelligent data-driven model based on EC combined with DL can adaptively change the communication range, lead to less interference, and increase the network throughput.

Figure 4.9 compares each algorithm's data transmission security performance under different numbers of end-users. It can be found that with the rise in end-users, the average delivery rate of network data shows an upward trend. The delivery rate of network data messages in the intelligent data-driven model is not less than 80% (Figure 4.9-a), the average leakage rate of network data is unchanged, and the data leakage rate is not more than 10% (Figure 4.9-b). Furthermore, the average delay decreases with the increase in the number of end-users, and the average delay of the intelligent data-driven model is stable at about 340 milliseconds (Figure 4.9-c). Therefore, the intelligent data-driven model with end-users in different amounts shows a significantly higher average delivery rate, lower average leakage rate, and shorter delay than the other algorithms, presenting brilliant data transmission security performance.

4.5 CONCLUSION

With the rapid development of wireless communication technology, the 5G network has achieved full coverage, and there is also a data explosion in wireless networks. This chapter combines EC with DCRFNN to construct a data-driven model of intelligent transportation IoT based on EC and DL. Finally, the performance analysis in simulation experiments indicates that the prediction accuracy of the intelligent data-driven model reaches 94.11%. The model achieves better data transmission security and lower transmission delay (about 340 milliseconds) than the other comparative algorithms. This model can provide a practical basis for the decision-making, management, and data transmission security of wireless networks in intelligent communication.

Still, there are some deficiencies in this chapter. Firstly, the data-driven model of intelligent transportation IoT based on EC and DL does not consider the diversity of service types in the existing network, including voice, data, and video. Therefore, a follow-up study will optimize the method to deploy MBSs intensively according to the actual application scenarios. Moreover, when the edge gateway processes the computing tasks unloaded from the terminal traffic generator, the priority of the processing tasks is not considered. Therefore, the priority of the tasks will be introduced in the future to perfect the edge gateway's deployment planning. Moreover, it is worth considering the workload required for traffic accident risk prediction and edge server computing resources and capacity, as well as studying the specific data placement strategy and task scheduling algorithm. These improvements are of great significance to the data-driven wireless network management in intelligent transportation.

REFERENCES

1. Huang, X., Yu, R., Pan, M., & Shu, L. (2018). Secure roadside unit hotspot against eavesdropping based traffic analysis in edge computing based Internet of Vehicles. IEEE Access, 6, 62371–62383.
2. Camacho, F., Cárdenas, C., & Muñoz, D. (2018). Emerging technologies and research challenges for intelligent transportation systems: 5G, HetNets, and SDN. International Journal on Interactive Design and Manufacturing (IJIDeM), 12(1), 327–335.
3. Duan, W., Gu, J., Wen, M., Zhang, G., Ji, Y., & Mumtaz, S. (2020). Emerging technologies for 5G-IoV networks: Applications, trends and opportunities. IEEE Network, 34(5), 283–289.
4. Yang, K., Shi, Y., Zhou, Y., Yang, Z., Fu, L., & Chen, W. (2020). Federated machine learning for intelligent IoT via reconfigurable intelligent surface. IEEE Network, 34(5), 16–22.
5. Song, X., Guo, Y., Li, N., & Zhang, L. (2021). Online traffic flow prediction for Edge computing-enhanced autonomous and connected vehicles. IEEE Transactions on Vehicular Technology, 70(3), 2101–2111.
6. Arena, F., Pau, G., & Severino, A. (2020). A review on IEEE 802.11 p for intelligent transportation systems. Journal of Sensor and Actuator Networks, 9(2), 22.

7. Zhang, J., & Letaief, K. B. (2019). Mobile edge intelligence and computing for the Internet of Vehicles. Proceedings of the IEEE, 108(2), 246–261.
8. Lv, Z., Lloret, J., & Song, H. (2021). Guest editorial software defined Internet of Vehicles. IEEE Transactions on Intelligent Transportation Systems, 22(6), 3504–3510.
9. Zhang, Y., Ma, X., Zhang, J., Hossain, M. S., Muhammad, G., & Amin, S. U. (2019). Edge intelligence in the cognitive Internet of Things: Improving sensitivity and inter-activity. IEEE Network, 33(3), 58–64.
10. Sodhro, A. H., Pirbhulal, S., Sodhro, G. H., Muzammal, M., Zongwei, L., Gurtov, A., ... & de Albuquerque, V. H. C. (2020). Towards 5G-enabled self adaptive green and reliable communication in intelligent transportation system. IEEE Transactions on Intelligent Transportation Systems, 22(8), 1–9.
11. Lv, Z., & Kumar, N. (2020). Software defined solutions for sensors in 6G/IoE. Computer Communications, 153, 42–47.
12. Ateeq, M., Habib, H., Afzal, M. K., Naeem, M., & Kim, S. W. (2021). Towards data-driven control of QoS in IoT: Unleashing the potential of diversified datasets. IEEE Access, 9, 146068–146081.
13. Chen, B., Wan, J., Celesti, A., Li, D., Abbas, H., & Zhang, Q. (2018). Edge computing in IoT-based manufacturing. IEEE Communications Magazine, 56(9), 103–109.
14. Li, X., Wan, J., Dai, H. N., Imran, M., Xia, M., & Celesti, A. (2019). A hybrid computing solution and resource scheduling strategy for edge computing in smart manufacturing. IEEE Transactions on Industrial Informatics, 15(7), 4225–4234.
15. Yang, S. R., Su, Y. J., Chang, Y. Y., & Hung, H. N. (2019). Short-term traffic prediction for edge computing-enhanced autonomous and connected cars. IEEE Transactions on Vehicular Technology, 68(4), 3140–3153.
16. Jaleel, A., Hassan, M. A., Mahmood, T., Ghani, M. U., & Rehman, A. U. (2020). Reducing congestion in an intelligent traffic system with collaborative and adaptive signaling on the edge. IEEE Access, 8, 205396–205410.
17. Arthurs, P., Gillam, L., Krause, P., Wang, N., Halder, K., & Mouzakitis, A. (2021). A taxonomy and survey of edge cloud computing for intelligent transportation systems and connected vehicles. IEEE Transactions on Intelligent Transportation Systems, 23(7), 1–16.
18. Shahbazi, Z., & Byun, Y. C. (2021). Improving Transactional data system based on an edge computing-blockchain-machine learning integrated framework. Processes, 9(1), 92.
19. Nallaperuma, D., Nawaratne, R., Bandaragoda, T., Adikari, A., Nguyen, S., Kempitiya, T., ... & Pothuhera, D. (2019). Online incremental machine learning platform for big data-driven smart traffic management. IEEE Transactions on Intelligent Transportation Systems, 20(12), 4679–4690.
20. Gu, Y., Lu, W., Xu, X., Qin, L., Shao, Z., & Zhang, H. (2019). An improved Bayesian combination model for short-term traffic prediction with deep learning. IEEE Transactions on Intelligent Transportation Systems, 21(3), 1332–1342.
21. Badawi, A., Chao, J., Lin, J., Mun, C. F., Jie, S. J., Tan, B. H. M., ... & Chandrasekhar, V. (2020). Towards the AlexNet moment for homomorphic encryption: HCNN, the first homomorphic CNN on encrypted data with GPUs. IEEE Transactions on Emerging Topics in Computing, 9(3), 1330–1343.
22. Lyu, F., Cheng, N., Zhu, H., Zhou, H., Xu, W., Li, M., & Shen, X. S. (2018). Intelligent context-aware communication paradigm design for IoVs based on data analytics. IEEE Network, 32(6), 74–82.
23. Zong, B., Fan, C., Wang, X., Duan, X., Wang, B., & Wang, J. (2019). 6G technologies: Key drivers, core requirements, system architectures, and enabling technologies. IEEE Vehicular Technology Magazine, 14(3), 18–27.

24. Luo, G., Li, J., Zhang, L., Yuan, Q., Liu, Z., & Yang, F. (2018). sdnMAC: A software-defined network inspired MAC protocol for cooperative safety in VANETs. IEEE Transactions on Intelligent Transportation Systems, 19(6), 2011–2024.
25. Autiosalo, J., Vepsäläinen, J., Viitala, R., & Tammi, K. (2019). A feature-based framework for structuring industrial digital twins. IEEE Access, 8, 1193–1208.
26. Xu, W., Zhou, H., Wu, H., Lyu, F., Cheng, N., & Shen, X. (2019). Intelligent link adaptation in 802.11 vehicular networks: Challenges and solutions. IEEE Communications Standards Magazine, 3(1), 12–18.
27. Liu, M., Yu, F. R., Teng, Y., Leung, V. C., & Song, M. (2018). Computation offloading and content caching in wireless blockchain networks with mobile edge computing. IEEE Transactions on Vehicular Technology, 67(11), 11008–11021.
28. Al-Kharasani, N. M., Zukarnain, Z. A., Subramaniam, S. K., & Hanapi, Z. M. (2020). An adaptive relay selection scheme for enhancing network stability in VANETs. IEEE Access, 8, 128757–128765.
29. Major, P., Li, G., Hildre, H. P., & Zhang, H. (2021). The use of a data-driven digital twin of a smart city: A case study of Ålesund, Norway. IEEE Instrumentation & Measurement Magazine, 24(7), 39–49.
30. Mitrentsis, G., & Lens, H. (2021). Data-driven dynamic models of active distribution networks using unsupervised learning techniques on field measurements. IEEE Transactions on Smart Grid, 12(4), 2952–2965.
31. Wang, Y., Yang, J., Liu, M., & Gui, G. (2020). LightAMC: Lightweight automatic modulation classification via deep learning and compressive sensing. IEEE Transactions on Vehicular Technology, 69(3), 3491–3495.
32. Wang, Y., Chen, Q., Gan, D., Yang, J., Kirschen, D. S., & Kang, C. (2018). Deep learning-based socio-demographic information identification from smart meter data. IEEE Transactions on Smart Grid, 10(3), 2593–2602.
33. Zhang D, Kabuka M R. Combining weather condition data to predict traffic flow: A GRU-based deep learning approach. IET Intelligent Transport Systems, 2018, 12(7): 578–585.
34. Xing, Y., Lv, C., Wang, H., Cao, D., Velenis, E., & Wang, F. Y. (2019). Driver activity recognition for intelligent vehicles: A deep learning approach. IEEE Transactions on Vehicular Technology, 68(6), 5379–5390.
35. Deepak, S., & Ameer, P. M. (2019). Brain tumor classification using deep CNN features via transfer learning. Computers in Biology and Medicine, 111, 103345.

5 Data-Driven Agriculture and Role of AI in Smart Farming

El Mehdi Ouafiq
SIRC- (LaGeS), Hassania School of Public Works,
Casablanca, Morocco

Rachid Saadane
SIRC- (LaGeS), Hassania School of Public Works,
Casablanca, Morocco

Abdellah Chehri
Department of Mathematics and Computer Science,
Royal Military College of Canada, Canada

CONTENTS

5.1 INTRODUCTION

Agriculture must change. The excessive use of phytosanitary products, soil tillage techniques, and production must adapt to current challenges [1]-[2]. The term "agriculture 4.0" was coined by reference to the buzzword "industry 4.0." The evolution of the industrial sector can be broken down into four stages. In the 19th century, the industrial sector was transformed with the help of steam engines: this was

DOI: 10.1201/9781003216971-6

uindustry 1.0. Then, the introduction of electricity on assembly lines corresponds to industry 2.0. The third phase of industrial transformation is characterized by informatics and the automation of processes on assembly lines. Finally, industry 4.0 brings together the latest changes thanks to digital technology and the possibility of interacting and communicating with different equipment. The agriculture industry has undergone the same changes [3]-[4].

Today, all machine tools and the entire production chain environment can transmit real-time information on their condition and performance. This information is centralized in the factory and makes it possible to control the various machines in consideration with the other devices' conditions. This makes it possible to automate and robotize a complete production chain comprising robots working on identical products simultaneously and in a coordinated manner. This saves time and, therefore, increases productivity [5].

This digitization of agriculture help farmers understand situations better and automate tasks that are thankless or difficult to perform because they are exact, fast, or repetitive. It will be possible, for example, to identify a weed in the process of emerging and control its growth with a weed-killer based on a robot that can work 24 hours a day [6–8].

Agricultural monitoring is an example where the Internet of Things (IoT) can help to increase productivity, efficiency, and output yield. Indeed, IoT in agriculture is a real "breaking wave." The miniaturization of electronic components, the marked improvement in energy autonomy, and the reduction in technology costs allow a considerable multiplication of sensors such as weather stations, soil moisture monitoring, or agricultural equipment monitoring. Recent work even suggests the development of sensors directly measuring the plant's physiological state (sap flow or water state of the crop).

In this chapter, using Oracle Database 11g, we propose a conceptual solution for the agricultural process at the data management level. The solution stores rainfall, temperature, and drought status in the raw format described by farmers. In addition, Excel files were used to fill in the crop statistical data. Furthermore, we propose an artificial intelligence (AI)-based architecture for smart farming. This architecture, called Smart Farming Oriented Big-Data Architecture (SFOBA), should be designed to guarantee the system's durability. Hence, the employment of data modeling to transform the business needs for smart farming into analytics. The goal is to provide a strategy for data ingestion that meets the data model while processing and storing the farm data, considering the data quality management and the system configuration.

The proposed architecture is built in layers to guarantee sustainability. It should handle the variety of sources and the large workloads considering the fault tolerance. Each data source has its particularity regarding data structure and types that should be handled at the data model's level. Intelligent farming analytics requires processing data in batch and real-time. The architecture should enable these capabilities, whether on-premises or a cloud platform, which allows farmers to have more flexibility while accessing the data (from smartphones, tablets, etc.) using only an Internet connection. The solution also requires having a single "source of truth" to all the intelligent farming analytics, the data lake (built-in layers). This solution can efficiently execute all technical controls and farming business rules. It also detects data anomalies and cleans them before storage. Data scientists and data visualization tools will also facilitate the farmers.

5.2 DATA MANAGEMENT AND INTEGRATION FOR TRANSFORMING SMART AGRICULTURE

5.2.1 DATA GENERATED BY MACHINE (DGM)

Today, a plethora of devices, sensors, and satellite imagery is used to collect agricultural data. This is data generated from machines (DGM) or IoT devices, including "smart" machines, satellites, drones, and intelligent robots [9]. This DGM mainly comes in via streaming, micro-batch, and small-batch data that has to be collected and processed in real-time or near-real-time applications. In the section that follows, we define the primary DGM sources. Also, it is essential to highlight the technical challenges related to collecting and processing this DGM and defining the data storage techniques. The primary objective is to store it in datasets so data analysts and data scientists can efficiently gather critical insights.

IoT and Wireless Connection: The global society is expected to be more digitized by 2030. Data will drive decisions and stimulate intelligent systems and processes, requiring near-instant and ubiquitous wireless connectivity [10].

With the IoT support of the sensor's interconnectivity, smart farming can benefit from numerous real-time monitoring and analysis opportunities. IoT-based solutions are deployed to help drive decisions in many areas of agriculture, mainly in water management and agriculture monitoring [11]-[12]. The data collection can be completed using networking standards such as RFID, LoWPAN, IEEE 802.15.x, 4G, and 5G. In addition, it is essential to note that the network interface can be used to share real-time data.

The term "smart system" or "intelligence system" includes the data collection, system processes, data analysis, automation, security, and logistics in intelligent farming. The IoT intervenes by offering smart devices, wireless media, seamless connectivity, and dictating the existence of machine-to-machine (M2M) communication [13].

The main objective behind using IoT in intelligent farming is interconnecting physical objects or devices through the Internet. As an example of IoT, components can be used for water management and intelligent irrigation analytics. We can also add temperature sensors, a pH sensor that measures the soil's nutrient content, or moisture sensors to measure the dielectric constant of the ground [14]-[15].

From a smart farming analytics perspective, the multiple DGM sources should be combined to collect the required data from the different sources [16]. Therefore, the system will connect the existing networks like satellites, mobile-cellular, and other yet-undefined networks, such as the 6G network [17].

The main idea is to remedy the restrictions of the current paradigm and technologies of the network by exploring a novel mechanism. The present scenario must support spatial distribution, water management, and equipment maintenance, to name a few. The paradigm implies adding new protocols, concepts, and architectures.

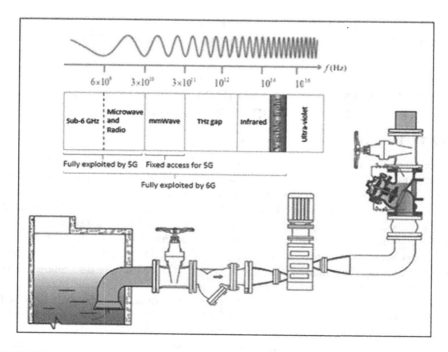

FIGURE 5.1   Development trend of communication technologies.

Many sensors should be deployed into factories, vehicles, lands, buildings, water systems, roads, etc. Thus, to support the enormous number of applications that should be integrated, proven wireless with high-speed communication should be investigated to process data-oriented activities [18]. This massive number of parameters is referred to as "big data requirements." Hence, to handle these requirements, a data migration strategy should be efficiently integrated, stimulating the need for a new approach in determining a dedicated big data architecture (SFOBA). This kind of architecture is proposed in this chapter.

Furthermore, from an intelligent farming analytics perspective, especially for collecting water distribution systems data, the new system, such as the 6G network, should integrate positioning, sensing, and communication to gather information from multiple sources (radar and biology communication included). Figure 5.1 shows an example of exploring a new system for water management.

Figure 5.2 represents the sensor's DGM used in this work. For the farm's mechanical system maintenance, we stimulated sensors installed on different farms' mining machines. Some critical parameters need to occur for insightful and errorless analytics to achieve accurate predictive industrial maintenance. The previous section mentioning Data Management Platform (DMP) contains the technical description of the intervention-related-parameters, next is the functional description:

- *Equipment_ Identifier (ID or Unique-Name):* Every physical sensor device has a unique identifier allowing device authentication, traceability, and analytics. Only one device will match the Equipment profile.

Column	Data Type
ORGANIZATION_ID	INT
PLAN_ID	INT
PLAN_NAME	CHAR
EQUIPMENT_ID	INT
MEASUREMENT_NAME	CHAR
MEASUREMENT_TYPE	CHAR
MEASUREMENT_RATE	FLOAT
MEASUREMENT_TIMESTAMP	TIMESTAMP
MEASUREMENT_TRESHOLD	INT
SECTION_ID	CHAR

Measurement Trend Description

FIGURE 5.2 Measurement trend description.

- *Organization_Identifier (ID or Unique-Name):* A field that identifies the organization responsible for carrying out the farm equipment measurements.
- *Section_Identifier (ID or Unique-Name):* A field that identifies the section of the farm (especially for large farms) where the equipment is located.
- *Inspection_Plan_Identifier (ID or Unique-Name):* Each inspection should be included on a plan associated with a specific measurement, equipment, organization, or section. From an audit perspective, this field is essential to perform analytics based on the plan failure (done mostly) and success.
- *Measurement_Name or Category:* e.g., Temperature, pH level, Vibration, Tension, etc.
- *Measurement_Type:* In smart farming, especially for mining extractions, the machines are giant and might have multiple sensors in different areas. For this perspective, a field that contains other modalities of the variable *Measurement_Name* should be added to get more details of the measurement (e.g., *Temperature_of_bearings, Temperature_of_engine*, etc).

5.2.2 DATA FROM HUMAN ORIGIN (DHO)

Data mainly generated from humans, or data from human origin (DHO), that is based on farmer experiences is a challenge from a data management perspective, because most of this type of data doesn't have a structure (or is semi-structured). Most of the time, the data will not be clean, requiring technical controls and verification to collect only the relevant data [19]. This DHO can be collected from social media, farmers, Excel files, etc.

The DHO is mainly considered one-time-load data for the data migration strategy, which contains some axis-of-analysis that can improve the analytics performed if processed, cleaned, and appropriately stored. In this work, we focused on DMP and DGM because DHO doesn't add any other new technical constraints than the ones required for the DMP and DGM processing.

5.3 DATA WISDOM IN SMART FARMING

From a smart farming perspective, we defined data wisdom as a combination of knowledge gathered from data analysis, data analytics, and predictive maintenance along with the farmer's experience-based knowledge (as shown here):

$$Data\ Wisdom = Data\ Analysis + Data\ Analytics + Predictive\ Maintenance \\ + Farmer's\ Experience$$

These data wisdom parameters are involved within big data techniques and are bound to big data technical challenges. A well-designed data model will facilitate the implementation of data gathering and enrichment techniques. The following sections define each of these parameters and how we exploit them in this research. We also describe our proposed solution in data modeling (from a farming business intelligence perspective) and the data governance requirements.

5.3.1 DATA ANALYSIS AND DATA ANALYTICS IN SMART FARMING

These big data techniques in smart farming should be well defined to avoid confusion, especially while using these terms to be operational in the way they are supposed to be. The data analysis in intelligent farming analyzes agriculture practices that have already happened. Statistical analysis uses mathematical inference to break down past agricultural processes into essential parts, identifying opportunities and solutions for changing them into better "smart" ones. A "farming" business intelligence analysis uses reporting and data visualization techniques to analyze and report historical farming business data to make tactical and strategic business decisions. It is mainly employed to exploit farm data to explain past events.

We consider the business intelligence analysis a preliminary step in intelligent farming analytics before performing predictive analytics. Business intelligence analysis relies on analyzing past data to extract meaningful insights and then creating appropriate models to inform farmers about their future productivity, operations, and equipment health. It is now clear that data analytics in intelligent farming are inspecting possible future events and also applies computational and logical reasoning to the analysis's results/parts.

Many attempts have been made to substitute farmers' perceptions by algorithms. Most of these efforts are taking advantage of big data technologies in different agriculture domains from which we can cite:

- Temperature monitoring: Collecting massive temperature data from IoT sensors and machines can be analyzed to provide valuable beneficial insights into equipment safety, irrigation, precipitation, and crop prices.
- Geographical segmentation: The goal is to inform farmers about their performances and why they have various performance levels based on geographical constraints since certain crops might have a higher level of concentration in a specific area and insignificant concentration in another [20]. In this research, we used these parameters to help farmers learn about crop concentration, yield statistics, and crop periods by doing spatial distribution analytics.
- Sustainability and water management: Numerous attempts [21–25] have been made to employ big data technologies and IoT-based systems in promoting watershed sustainability. The next section proposes data modeling techniques that make water management analytics queries operational and practical on top of big data architecture.
- Bioinformatics can quickly process, store, analyze, and visualize biological data. The perspective in this research is to provide a data architecture that makes biological data available in electronic form on a dedicated "specific" source, which will lessen the time and the efforts of data collection from bioinformatic primary and secondary databases and various literature sources.

With the development of open-source big data technologies and the rapid growth of big data communities, due to the vast demand from multiple industries for various types of businesses, different big data techniques can remedy the traditional decision-making systems limitations [26].

This work links agricultural information gathered mainly from IoT systems and relational database management system (RDBMS) with big data/data-science techniques and tools to assemble insights by performing smart farming analysis, analytics, and predictive maintenance.

To schematize, the agricultural-based insights fall into three categories: Spatial distribution-based insights, water management-based insights, and equipment maintenance-based insights.

a. *Spatial Distribution:* This category should include any smart farming analytics that allow farmers to have crop distribution (e.g., "best crop to be grown" or any factor contributing to productivity). This work has chosen the prediction of "drought-affected regions," which might also impact crop productivity. From a "farming" business intelligence perspective, the spatial distribution analytics should include three primary epics (at least one dashboard). Table 5.1 demonstrates in detail the Key Performance Indicators (KPIs) of each epic:

- *Climate Overview:* This helps to have preliminary "patterns" and parameters that facilitate the comprehension of the productivity results.
- *Productivity:* Studies the production and the productivity of the farm.
- *Global issues:* What is impacting the crop distribution.

From a data science perspective, the spatial distribution analytics should enable farmers to have visibility on future events, as shown below:

- *A climate overview* should enable farmers to compare the existing patterns with the required and predicted ones.
- *The productivity* dashboard should enable farmers to label the productivity per crop per season from very-low, low, medium, to high and find correlations between the rainfall amount (or any other relevant parameter, e.g., temperature) and the productivity.
- The *global issues* dashboard should enable the farmer to label the problem (or the issue) level from low, medium, to high. This parameter also shows the impact on crop production, allowing the farmer to predict the best crop to be grown in a specific area during a particular time.

From a data visualization perspective, all the various data sources in agriculture cause the data to become massy and more compound due to the horizontal scalability of the big data architectures. The data are now ready to use visualization techniques to form knowledge-discovery and farming business intelligence, which is crucial to finding patterns, especially in semi-structured and unstructured farming data [28]. The critical issue now is how to create meaningful, integrated, and clear visualizations.

The visualization is a front-line tool to disclose the detailed "farming" and "business" process, such as obtaining real-time farmland information, which is an advanced method for the data-modeling presentation (which will be described in the upcoming section). In this research, we compared different data visualization "graphs" to present what best fits the intelligent farming analytics.

The bar chart entails rectangular chunks (or blocks) of various of heights, where the vertical axis reveals values (in mm, Pa, C°, etc.) of temperature, rainfall, pressure, and productivity with an extensive range, e.g., values/frequencies, total, unit-of-measurement, counts, and percentages. The horizontal axis indicates the specific category, whether it is a geographical dimension (e.g., district), a time dimension (e.g., date, month, and season), or an "indicator" of a specific measure (e.g., temperature, rainfall).

For smart farming-based insights, as a best practice, the bar chart provides better visibility when measuring the change over time (as shown in Table 5.1), especially from a farming-analysis perspective. Moreover, the stacked bar chart in its different forms (inverting, diverging, and classical) can be the most convenient visualization method to present multiple attributes on the data.

The line chart plots single points on the two horizontal/vertical axes and joins up the adjacent points by straight lines. The vertical axis can contain the measurements of drought status and equipment measurement trends, such as vibration, temperature, attrition, and tension (as shown in Table 5.2), while the horizontal axis conventionally represents a time dimension (e.g., month, date, etc.).

TABLE 5.1

Key Performance Indicators (KPIs) and Charts of the Relevant Epics from a Business-Intelligence and Data-Science Point of View

Epic	KPI	Description	Data Visualization Chart
Climate Overview	Min and Max	Visualizing the Min and Max of the rainfall, temperature, and pressure	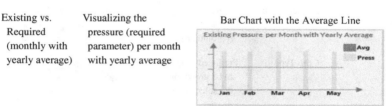
	Existing vs. Required (monthly)	Visualizing the existing and required precipitation and temperature (relevant parameters) per month	Bar Chart
	Existing vs. Required (monthly with yearly average)	Visualizing the pressure (required parameter) per month with yearly average	Bar Chart with the Average Line
	Drought Status	Visualizing the drought status per month	Line Chart
Productivity	Min and Max	Visualizing the Min and Max productivity per a specific district	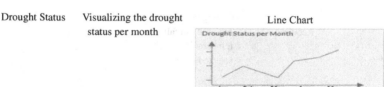
	Top 3 districts	Visualizing the Top 3 districts based on the total production	Bar Chart with the Productivity Line

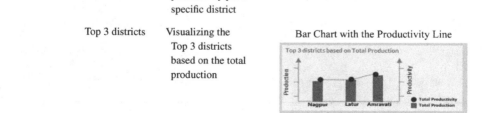

(Continued)

TABLE 5.1 *(Continued)*

Key Performance Indicators (KPIs) and Charts of the Relevant Epics from a Business-Intelligence and Data-Science Point of View

Epic	KPI	Description	Data Visualization Chart
	Crop Positioning	Productivity and rainfall correlational	Perceptual Map 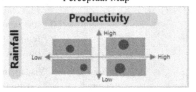
Drought and Crop Distribution	Value and Average	Recorded temperature, pressure, and rainfall along with their averages	
	Drought Status	Predicted drought level on the selected region/district	
	Crop Type	Predicted crop type to be grown on the selected region/ district	
	Existing vs. Required	Existing and required rainfall and temperature on the selected month	Bar Chart 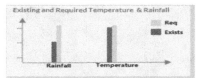
	Report	Drought and crop distribution general report	Table

District	Amravati
Year	2011
Month	June
Rec. Rainfall	58mm
Avg. Rainfall	101mm
Rec. Temp	34 C°
Avg. Temp	32 C°
Rec. Pressure	1000
Avg. Pressure	1004
Drought Classification	High

TABLE 5.2

KPIs and Charts of the Relevant Epics from an "Industrial" Mechanical System Maintenance Perspective

Epic	KPI	Description	Data Visualization Chart
Measurement Trend – Logistic Organization	Measurement trend categories	Visualizing the measurement trend categories per a selected organization for a specific date	Measurement Category per Selected Organization Temp / Vibr / Tens
Measurement Trend – Equipment	Vibration/ Temperature/ Tension/Attrition Measurement Trends	One KPI for each visualization of the measurement trend for the critical categories (Vibration, Temperature, Tension, Attrition) per equipment, per a specific measurement type, per a specific date (or month)	Line Chart Vibration Measurement Trends
Interventions	Measuring Interventions	One KPI for each of the visualizations of the number of interventions per Equipment/ Organization/Section by measurement category and type. Colored differently are the measurements Nbr <10, Nbr ≤30, Nbr ≥30.	Bar Chart Nbr of Interventions per Equipment Nbr >= 30 / Nbr <= 30 / Nbr < 10
Predictive Maintenance	Intervention Report	A report of interventions that indicates/predicts to the farmers if a machine will fail; this table should include the equipment, organization, section, timestamp, and prediction	
	Decision Tree	A decision tree to have clear visibility on the decision rule that predicts if an intervention is needed	

The continual line implicates a tendency over time. Therefore, it is advantageous to use a line chart in farming analysis to show trends over time, e.g., tracking equipment behaviors, drought status, or any global problem. In addition, the combined bar-line chart can be used to provide more consistent visualization for multiple measurements with the same dimension.

The perceptual map in smart farming (especially in spatial distribution) represents the farmer's perceptions about particular attributes, such as productivity based on an organization, crop type, or temperature axis of analysis.

The pie chart exhibits the proportions of a whole. Therefore, it can be effective only if the portions are essential and more significant (or substantially different). Nevertheless, it is effective in farming mechanical system maintenance to compare each quantity (represented as a category) of the measurement trends. Especially that temperature, vibration, and tension types can have many measurement trends that differentiate them from the other portions.

> b. *Water Management:* Water leaks, especially in water-distributed systems, can be reduced by maintaining inappropriate water pressure, which helps predict suggested emergency actions. Furthermore, using water distribution systems data collected from IoT-Devices, we can develop dedicated queries to analyze consumption patterns and visualize them in dashboards [29].

Also, for the smart irrigation perspective, big data analytics is heavily involved in automatically analyzing the soil-based data and providing adequate and automatic irrigation based on smart water dripping. The soil-based data, such as moisture, temperature, and pH can be ingested in the data lake and used to analyze the enrichment layer to drive an automatic decision on whether the land should be irrigated [30].

Table 5.3 describes the role of each sensor, IoT component, data, and its usefulness on the analytics from an intelligent water-dripping perspective.

IoT sensors like iron analyzer, consumption sensor, water leak detector, and granular activated carbon filter (GAC) detector can also provide decisive data for water consumption, accumulation of iron deposits on ductile iron pipes, and the total amount of total trihalomethane (TTHM) in the water, etc. If that information is collected, processed, and analyzed in real-time, it can provide alerts (consolidated based on KPIs and visualized on dashboard/reports), which help farmers make a timely and correct decision to prevent loss.

5.4   DATA MODELING FOR SMART FARMING

The success of data-driven analytics for smart farming is related to how organized the data management is. This, in turn, means developing actions, policies, and architectures to manage the fundamental requirements of the data life cycle of the farm in a proper manner and align the technologies with the farming business needs. The development of data architecture for smart farming involves accurate rules, data models, data storage, data processing, and integration. Therefore, a

TABLE 5.3

Sensors/IoT components, Data, and Analytics for Smart Water Dripping

Sensor/IoT Component	Data	Analytics
pH Sensors	Nutrient content	Measuring and analyzing the content of nutrients in the soil and defining the amount required for irrigation
Temperature Sensor (e.g., RTD)	Temperature	Analyzing the resistance to measure the temperature (based on a correlation, more resistance refers to more temperature)
Moisture Sensor	Dielectric constant and resistance	Analyzing the dielectric constant and the resistance to measure the moisture content (more resistance refers to moister content; a higher level of dielectric content refers to a higher moisture content level)
Water Pump		Will operate based on the analysis result on the intelligent farm's device data. As a result, the land will be supplied with water with an analysis-based given quantity.

- Sensors can detect the content level of the water in the soil
- The sensors mentioned transmit the data into the data lake to enrich the data with a dedicated calculation and estimate the precise required quantity of water
- The soil and crop parameters can also be used to enrich the data to control the subsequent watering of plants
- The status of irrigation can be continuously updated at the data lake level (using big data technologies)
- Algorithms like naive Bayes can be used to suggest fertilizers
- Weather forecasting can be used to estimate the amount of upcoming rain
- Finally, the crops (on different areas/lands) are going to be watered accordingly (based on the above parameters), e.g., the system should be able to automatically adjust/reduce the water that will be supplied to the crops if heavy rainfall is coming (predicted)

well-designed data model should be built to define and analyze the data requirements. This includes:

1. The farm databases/sources
2. The relationships between the data items
3. Data constraints
4. Conceptual representation of the data structure

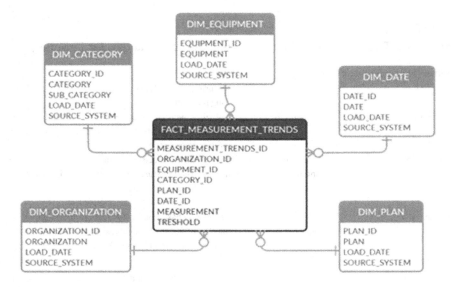

FIGURE 5.3   Star schema for measurement trends of the mechanical system.

From an intelligent farming perspective, the star schema will consider the farming processes as fact, e.g., spatial distribution, water management, or mechanical system maintenance, and dimensions, e.g., geography, time, crop-type, etc. Ordinarily, in this case, the star schema will contain these two levels, a simplified design for smart farming analytics, as shown in Figure 5.3.

But for spatial distribution, the agriculture season is an essential axis of analysis that requires more granularity of the time dimension. At the same time, especially for large-farming enterprises, they might have sections in different continents, countries, states, and cities. We have to study the climate, crop production, productivity, and crop distribution in every district, which necessitate descending into a significantly lower level of granularity of the dimension geography. This category of arrangement/configuration is referred to as snowflake schema, as described in Figure 5.4.

The historical data is also an essential factor for the success of the analysis and prediction-based analytics. These data allow the farmers to track their processes, transactions, equipment, and the historical water consumption data to predict future consumption.

The star schema has a few limitations when dealing with historical "big" data because it requires highly cleansed data. However, the star schema doesn't scale well when upstream systems change. Here, the data vault 2.0 comes into the picture that shows better performance than 3NF and dimensional modeling to create historical data for smart farming analytics and master data management.

The data vault 2.0 is considered a business intelligence system that offers a massively parallel architecture for big data, unstructured data, and real-time processing through data warehousing. The constituent that incorporates a data model can deal with multi-latency and cross-platform data persistence, including disciplined agile

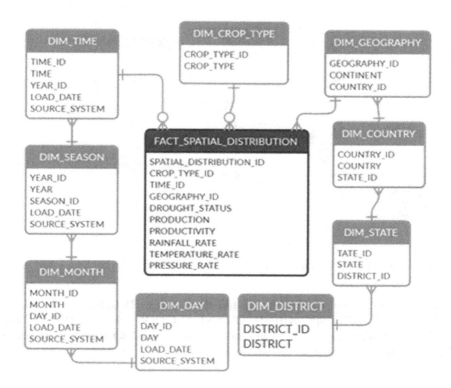

FIGURE 5.4 Snowflake schema for the spatial distribution.

deliveries (based on agile methodology). It is also important to mention that a star schema can be built from a business data vault.

The data vault has three main types of tables:

- *HUB:* Contains a business key list, each one has a surrogate key, along with metadata describing the source system, the origin of the business keys, and the time
- *LINK:* To set up relationships between these business keys (ordinarily it describes a relationship of many-to-many), these links are essential to deal with granularity changes and lessen the impact of appending a new business key to a "linked" HUB
- *SAT/Satellites:* Similar to Ralph-Kimball SCD type II, the SAT carries the descriptive attributes and can change over time. The LINKs and HUBs form the structure of the data model. The SATs include temporary descriptive attributes and the metadata that link them to the parent LINK or HUB table. This mechanism gives strong historical capabilities by authorizing queries to go back in time, based on the metadata attributes inside an SAT table that contains a date where the record becomes valid and another date where it is expired.

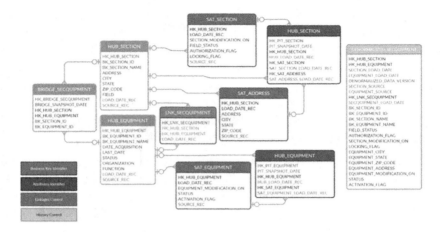

FIGURE 5.5   Data vault for historical data.

The data vault model consists of the following tables: HUB_SECTION, HUB_ EQUIPMENT, LNK_SECQUIPMENT, SAT_SECTION, SAT_ADDRESS, and SAT_EQUIPMEN. Figure 5.5 represents how we can use data vault to maintain historical data about each section's farm equipment.

5.5   DATA MIGRATION STRATEGY FOR SMART FARMING

The IoT is heavily involved in most of the smart farming architecture. The IoT provides real-time monitoring, coverage, and connectivity. In addition, the IoT network allows multiple objects to be connected over all areas the farm. It is also necessary to remember that traditional storage and processing decisional systems that scale vertically proved incapable of handling various data sources (DMPs, DGMs and DHOs). This is mainly because the amount of data will be increased with a high velocity and with an unexpected volume and quality of the data.

These "*variety,*" "*velocity,*" and "*volume*" requirements represent what is so-called "big data processing." This work proposes a data migration strategy and uses a big data architecture. The model enables us to exploit the horizontal scalability to fully use all the valuable farms data, which is not supposed to be stored on a single node. This farming data will be too large (with an incredible velocity). Therefore, it should be distributed beyond multiple nodes.

5.5.1   Hadoop Architecture

Thanks to the MapReduce framework of the Hadoop ecosystem, the massive farm data can be distributed and replicated across multiple nodes, processed in parallel, modeled, and stored. This will guarantee high availability and more scalability than traditional decision-making systems. Hadoop has three main components:

- *HDFS:* A Hadoop distributed file system
- *YARN:* A distributed operating system
- *MapReduce:* A framework for writing and processing data in a Hadoop environment

Hadoop can be built and configured on commodity hardware where a group of machines is connected via a network. Furthermore, and because of the horizontal scalability, the hardware components in Hadoop can be maintained easily (and are less costly).

Hadoop is based on a master-worker node architecture, where each node has storage and processing part. The main components of this architecture are:

- *Resource Manager:* For each cluster, a resource manager receives and runs the applications—one of its prominent roles in handling resource allocation and keeping track of the total resources and nodes
- *Master:* Takes responsibility for managing the execution of the MapReduce jobs (for each job, a worker node will act as application master)
- *Job History Server:* Provides information about the completed job.
- *Node Manager:* For each data node, there is a node manager. This node replaced the Task Tracker in Hadoop 1.x, which manages resources and deployment on a node.
- *Name Node:* Responsible for making all the decisions about block replication. It is responsible for managing the file system metadata, monitoring block replication, performing operations (e.g., close, open, rename a file) related to the file system namespace, and determining the mapping between the blocks and data nodes. There is a secondary name node that can be optional, but it is very important for guaranteeing high availability when the name node is down.

The MapReduce framework supports task scheduling, monitors them, and reruns failed tasks. A MapReduce job splits its input dataset into blocks processed by map tasks in a parallel fashion. The outputs of the map jobs will be sorted and used as input for the reduced tasks.

5.5.2 Technical Constraints

From a smart farming analytics perspective, the quality of this data cannot be expected, which will significantly impact the accuracy of the analytics. Therefore, handling data quality should be one of the top priorities when building the data architecture for smart farming.

The DGMs come in stream data and small batches (e.g., water consumption, temperature, and mechanical measurement). It is also essential to consider that some small files can't be combined into one large file, like the large corpus of satellite/drone images used for crop monitoring.

Each image is a separate file. Every block, directory, and file in Hadoop Distributed File system (HDFS) is represented as a 150-byte object in the name node's memory.

A considerable number of small files are supposed to be generated in a smart farm, which implies handling a massive amount of metadata that might make the name node run out of memory. On the other hand, an archiving concept called Hadoop Archives (HAR) can be used in HDFS to pack multiple files into a smaller amount of HDFS files. The problem with this HAR concept is the need for two index file reads and the data file read, which slows down the reading of files.

As an alternative to the master-index approach, New Hadoop Archive architecture was proposed to employ a single-level index, based on a hash table containing the index information to split them over multiple index files. The New Hadoop Archive concept outperforms HAR by 85.47% while accessing 80,000 files [34]. Still, the HDFS raw file provides better computation at reading, which leads us to another constraint.

The computation performance in the Hadoop environment has been the subject of much research. For example, Rattanaopas et al. [31] has compared multiple compression algorithms to improve computation performance and increase the storage space while compressing large-small files in a Hadoop environment. On a WordCount example, bzip2 outperformed the other algorithms gzip, snappy, and LZ4 in terms of computation performance and saved 70% more storage space than a raw HDFS text file (which still shows better computation performance).

Here is one essential factor: Hadoop's job scheduling and resources management can significantly impact performance if they are not appropriately developed. A scheduler defined the jobs that should be running (when and where) and the allocated resources for each job. There are a few schedulers that are supported by MapReduce and YARN from which we can cite:

- *FIFO:* The allocation of the resources is completely based on the time of the arrival
- *Fair*: The weighted pools where the resources will be allocated, based on a fair-sharing on each pool
- *Capacity:* FIFO-based scheduling on each pool to allocate the resources. A job scheduling algorithm in a Hadoop environment should handle the data block migration, resources, and network constraints at the first level

5.6 SMART FARMING ORIENTED BIG DATA ARCHITECTURE

Hadoop solutions for intelligent farming analytics can be effective alongside a data migration strategy. That intercalates data modeling, data processing, data storage, data governance, data quality, and the administration/configuration of the Hadoop system on top of a data architecture that meets all these requirements and answers all the technical constraints. However, the proposed solutions employed Hadoop technologies in agriculture to handle the massive and various farm data and relieve the limitations of traditional decision-making systems. The Hadoop solutions by choosing a couple of clusters, grouping them, and then installing the needed tools for the analytics might or might not be profitable as a solution, especially for the sustainability of the innovative farming analytics system.

Given the fact that a SFOBA should be designed to guarantee the system's durability, the employment of data modeling can transform the business needs for smart farming into analytics. The goal is to provide a strategy for data ingestion that meets the data model while processing and storing the farm data, considering the data quality management and the system configuration.

In the field of big data, there are three main types of architecture from which our SFOBA was inspired:

- *Data Lake:* Designated for processing and storing large volumes of data mainly in batch, where raw data can be in different formats: structured, semi-structured or unstructured. This architecture comprises four layers:
  - *(1) Data sources* (e.g., files, web services, logs, etc.)
  - *(2) Ingestion system:* Workflow for data ingestion based on Spark, Sqoop, Kafka, and other technologies
  - *(3) Data Lake:* The Hadoop Data Warehouse where the data will be stored
  - *(4) The layer of Interrogation:* Performed based on Hive SQL, Spark SQL, HBase, etc.
- *Lambda Architecture*: Aims to handle batch processing within data blocks and real-time processing continually depending on the data flow. This architecture is not mainly made for data storage. Still, it focuses on joining the massive batch data stored on the batch layer with the streaming data from the real-time/speed layer that will be calculated or incremented on dedicated views. Finally, the serving layer is where the client will have direct access to the needed views for the analytics
- *Kappa Architecture:* Since the Lambda architecture is a bit complex because of the merge between the batch and streaming layers, the Kappa architecture was designed for data processing only without storing the data permanently. This architecture considers: (1) Data source layer; (2) the real-time layer where data pipelines will build generally using Kafka to store the current messages that will be processed Spark, Storm, or other technologies; (3) the layer of views, which is the output of the data processing; and (4) the service layer where data can be queried using Hive, HBase, Cassandra, etc.

A recent solution called "Delta Lake" was introduced by Databricks that answers many problems cited in the literature, especially in big data. It is a flow management mechanism of the data pipeline from data sources to the data lake and vice versa and is based on the Databricks File System (DBFS) and Apache Spark to provide a transactional storage. The main idea is to provide an abstraction layer of the data lake that relays on a Spark-optimized table that supports Parquet-type files and manages logs that monitor the table's changes. The main challenges with Delta Lake are:

- Ensuring the schema enforcement while introducing a new table
- Repairing the newly created tables (mainly hive tables)
- Refreshing the metadata frequently (to avoid having the data ingested without being able to query on it because of the metadata updates)
- Traffic of small files, especially for distributed computations

- The struggle of sorting the data by index (e.g., an I.D.) where data is partitioned (e.g., by time) and distributed across multiple files. This Delta Lake solution provides many advantages, especially for ACID transactions. Many jobs can write and modify the target table (or dataset) simultaneously and get a consistent view. Also, the reading jobs will not interfere while writing/revising the dataset. And the automatic file management accelerates the data access by grouping the data into larger files that can be read effectively.

We took the best of these techniques in the solution and adapted them for smart farming analytics to prevent our data lake from drowning in the massive farm data.

The SFOBA, as shown in Figure 5.6, which Lambda inspired because the *query = function(farming-physical-data-model)*, is built based on five horizontal layers (described next). Each has its functionalities and performs specific write and read actions on the previous layer. This architecture allows for more agility, flexibility in security maintenance, and management for different formats and historical data [32].

1. The first layer is a *shared area* that can also be considered a landing area built on top of a network file system (NFS), containing a structured folder based on the farming business logic. This shared area is not a schema, but a group of folders on NFS that are oriented to save data temporarily. Thus, this shared area should have more drivers and more space for storage. In most cases, this area will be used as a gateway of the Hadoop platform, a destination of the manual files (mostly DHOs and sometimes DMPs), and a landing area of the data coming from less secured sources (mostly DGMs).

2. The second layer is a *raw zone* where the data will land directly from the source into the Hadoop platform. The data will be directed on specific directories (based on the physical data model). The metadata will be assigned to it in Hive Metastore to be affected by an external table. In this area, considering the storage space and the computation performance, while importing DMPs mainly from RDBMS using Sqoop jobs, the compression algorithms can be employed to write the inputs using bzip2 as recommended. To maintain the security of the user privileges and prevent conflicts between the MapReduce jobs that might lock the tables in this raw zone, only the data engineers (and the data owner) can access this layer.

3. The third layer is a *structured layer* where the raw data will be stored based on a structure and specific data types, unlike the raw zone where all the data will be stored as character data. In this layer, technical controls and farming business rules verifications will be launched to check the data quality, detect anomalies, and inform the data steward (through emails and dashboards) as described on the activity diagram of Figure 5.7. Also, the historical data will be stored based on the data vault model. Only the data engineers, data stewards (and the data owner) should access this layer.

4. The fourth layer is a *trusted zone* where the data will be stored based on a data model (e.g., Star Schema, Snowflake, or Fact-Constellation) after checking the data quality and cleaning it on the previous *structured layer*.

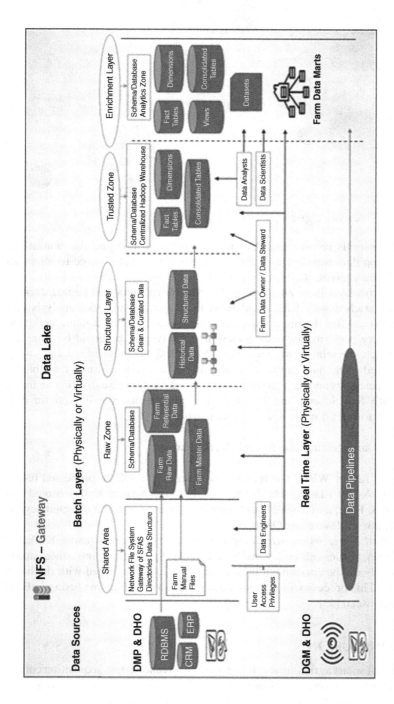

FIGURE 5.6 Smart-farming-oriented big data architecture.

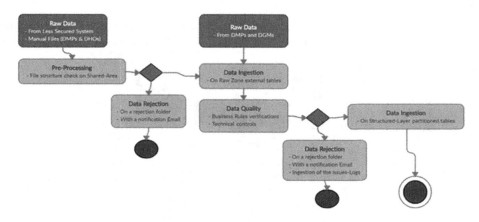

FIGURE 5.7   Activity diagram for data quality checks.

This zone is recommended as the "source of truth" and the centralized Hadoop data warehouse of the smart farm, the single source to all smart farming analytics KPIs.

5. The fifth layer is an *enrichment layer* where the data will be transformed and enriched with KPI calculations for a specific analytics/analysis perspective. After the ingestion success on this zone, from a name node and memory performance management perspective, a job should be launched to delete the related raw zone files (to decrease the metadata that has to be handled). The data in this layer should be stored in a format readable by data scientists and data analysts and suitable for data visualization analysis. Mostly Star Schema and Snowflake logical data models will store the data in the form of domain-specific data marts.

The SFOBA architecture also has two vertical layers:

- *Batch Layer:* Where batch and micro-batch data will be processed (using technologies like Apache Spark and Python, Apache Kylin), stored (on HDFS, Hive, and Impala), and automated in the form of a directed acyclic graph workflow (using Oozie or Airflow).
- 2. *Real Time:* where streaming data will be consumed (using Kafka or Flume), processed, and transformed (using SparkStreaming), then stored directly on the enrichment layer (using HBase) to be joined with the batch data. This process will be automated within a data pipeline (using NiFi or Streamsets) [33].

5.7 CONCLUSION

The future of smart agriculture will be based on advanced data acquisition combined with several technologies – all of which involve data management. Given the fact, and as we described in this chapter, a smart data management can provide many solutions to facilitate the farmers' work when dealing with their daily processes.

Therefore, we proposed an innovative farming-oriented big data architecture, SFOBA, that provides a platform to manage farming data in different layers, executing tasks accordingly, allowing more flexibility and agility.

In this chapter, we have proposed some of the leading technical constraints, a data quality process, a way to model the farm's data based on the business logic, and a comparison between machine learning algorithms where the ID3 gave the highest level of accuracy.

In the future, an abstraction layer can be built on top of the SFOBA that automatically handles the technical constraints based on rules defined on the system, without adding technical constraints as actions on the directed-acyclic graph of the ingestion workflows. The project management aspect, DataOps methodology, and the Agile method give an enterprise tincture to the smart farm data management.

The ultimate goal of our research is to ensure agricultural sustainability and environmental protection, secure energy-consumption, and reach a reasonable level of quality, productivity, and high efficiency.

REFERENCES

1. The State of Food and Agriculture 2014: Innovation in Family Farming 2014, Food and Agriculture Organization of the United Nations. Available online at: http://www.fao.org/publications/sofa/2014/en/
2. This Is How to Sustainably Feed 10 Billion People by 2050, World Economic Forum. Available online at: https://www.weforum.org/agenda/2018/12/how-to-sustainably-feed-10-billion-people-by-2050-in-21-charts/ (Accessed on 5 December 2020).
3. H. Chung, D. Kim, S. Lee and S. Cho, "Smart Farming Education Service Based on u-learning Environment," 2019 21st International Conference on Advanced Communication Technology (ICACT), Pyeong Chang Kwangwoon, Korea (South), 2019, pp. 471–474, doi: 10.23919/ICACT.2019.8701949.
4. N. Islam, B. Ray and F. Pasandideh, "IoT Based Smart Farming: Are the LPWAN Technologies Suitable for Remote Communication?" 2020 IEEE International Conference on Smart Internet of Things (Smart-IoT), Beijing, China, 2020, pp. 270–276.
5. D. Glaroudis, A. Iossifides and P. Chatzimisios, "Survey comparison and research challenges of IoT application protocols for smart farming", Computer Networks, vol. 168, 2020.
6. X. Feng, F. Yan and X. Liu, "Study of wireless communication technologies on Internet of Things for precision agriculture", Wireless Personal Communications, vol. 108, no. 3, pp. 1785–1802, 2019.
7. X. Shi, X. An, Q. Zhao, H. Liu, L. Xia, X. Sun, et al., "State-of-the-art internet of things in protected agriculture", Sensors, vol. 19, no. 8, pp. 1833, 2019.
8. E. M. Ouafiq, A. Elrharras, R. Saadane, M. E. Aroussi, A. Chehri, "IoT in Smart Farming Analytics, Big Data Based Architecture," 13th International Conference on Human Centred Intelligent Systems (HCIS-20), Split, Croatia, June 2020.
9. M. Bacco, A. Berton, E. Ferro, C. Gennaro, A. Gotta, S. Matteoli, et al., "Smart Farming: Opportunities Challenges and Technology Enablers" in 2018 IoT Vertical and Topical Summit on Agriculture-Tuscany (IOT Tuscany), IEEE, pp. 1–6, 2018.
10. A. Khanna and S. Kaur, "Evolution of Internet of Things (IoT) and its significant impact in the field of precision agriculture", Computers and Electronics in Agriculture, vol. 157, pp. 218–231, 2019.
11. O. Elijah, T. A. Rahman, I. Orikumhi, C. Y. Leow and M. N. Hindia, "An overview of Internet of Things (IoT) and data analytics in agriculture: Benefits and challenges", IEEE Internet of Things Journal, vol. 5, no. 5, pp. 3758–3773, 2018.

12. V. Y. Chandrappa, B. Ray, N. Ashwath and P. Shrestha, "Application of Internet of Things (IoT) to Develop a Smart Watering System for Cairns Parklands – A Case Study", 2020 IEEE Region 10 Symposium (TENSYMP), pp. 1118–1122, 5–7 June, 2020.

13. X. Yang et al., "A survey on smart agriculture: Development modes, technologies, and security and privacy challenges," in IEEE/CAA Journal of Automatica Sinica, doi: 10.1109/JAS.2020.1003536.

14. A. Łukowska, P. Tomaszuk, K. Dzierżek and Ł. Magnuszewski, "Soil Sampling Mobile Platform for AGRICULTURE 4.0," 2019 20th International Carpathian Control Conference, Poland, 2019, pp. 1–4.

15. K. Anand, C. Jayakumar, M. Muthu and S. Amirneni, "Automatic Drip Irrigation System using Fuzzy Logic and Mobile Technology", Conf. on Techn. Innovations in ICT for Agriculture and Rural Development, 10–12 July, 2015.

16. L. G. Paucar, A. R. Diaz, F. Viani, F. Robol, A. Polo and A. Massa, "Decision Support for Smart Irrigation by Means of Wireless Distributed Sensors", IEEE 15th Mediterranean Microwave Symposium (MMS), Lecce, Italy, 2015.

17. V. Ziegler and S. Yrjola, "6G Indicators of Value and Performance," 2020 2nd 6G Wireless Summit (6G SUMMIT), Levi, Finland, 2020, pp. 1–5.

18. S. Mathivanan and P. Jayagopal, "A big data virtualization role in agriculture: A comprehensive review", Walailak Journal of Science and Techology, vol. 16, no. 2, pp. 55–70, 2018.

19. O. Debauche, et al., "Data management and internet of things: A methodological review in smart farming", Internet of Things, vol. 14, p. 100378, 2021.

20. S. Senthilvadivu, S.V Kiran, S. P. Devi and S. Manivannanc: "Big Data Analysis on Geographical Segmentations and Re-source Constrained Scheduling of Production of Agricultural Commodities for Better Yield," In: Fourth International Conference on Recent Trends in Computer Science & Engineering, Elsevier, vol. 87, pp. 80–85, 2016.

21. J. Dela Cruz, R. Baldovino, A. Bandala and E. Dadios, "Water Usage Optimization of Smart Farm Automated Irrigation System Using Artificial Neural Network", Fifth International Conference on Information and Communication Technology (ICoICT), Malacca, Malaysia, 17–19 May, 2017.

22. A. Y. Hoekstra and P. Q. Hung. Virtual water trade: A quantification of virtual water flows between nations in relation to crop trade. Value of Water Research Report Series. n.11, United Nations Educational, Scientific and Cultural Organization–Institute for Water Education, 2020.

23. H. H. Kadar, S. S. Sameon and P. A. Rafee, "Sustainable Water Resource Management Using IOT Solution for Agriculture," 2019 9th IEEE International Conference on Control System, Computing and Engineering (ICCSCE), Penang, Malaysia, 2019, pp. 121–125.

24. Y. Liu, Y. Zhang, Y. Chen and J. Tong, "Research Based on Real Time Monitoring System of Digitized Agricultural Water Supply," 2011 6th International Conference on Computer Science & Education (ICCSE), Singapore, 2011, pp. 380–383.

25. A. Saad, A. E. H. Benyamina and A. Gamatié, "Water management in agriculture: A survey on current challenges and technological solutions," in IEEE Access, vol. 8, pp. 38082–38097, 2020.

26. W. A. Goya, M. R. D. Andrade, A. C. Zucchi, N. M. Gonzalez, R. D. F. Pereira, K. Langona, T. C. M. D. B. Carvalho, J. E. Mångs and A. Sefidcon. "The Use of Distributed Processing and Cloud Computing in Agricultural Decision-Making Support Systems," 7th International Conference on Cloud Computing, USA, 2014, p. 721–8.

28. Z. Ünal, "Smart farming becomes even smarter with deep learning—A bibliographical analysis," in IEEE Access, vol. 8, pp. 105587–105609, 2020,

29. H. Y. El Sayed, M. Al-Kady and Y. Siddik, "Management of Smart Water Treatment Plant using IoT Cloud Services," 2019 International Conference on Smart Applications, Communications and Networking (SmartNets), 2019, pp. 1–5.
30. A. Slalmi, H. Chaibi, R. Saadane, A. Chehri, G. Jeon, H. K. Aroussi, "Energy-efficient and self-organizing Internet of Things networks for soil monitoring in smart farming," Computers & Electrical Engineering, vol. 92, 2021.
31. K. Rattanaopas and S. Kaewkeeree, "Improving Hadoop MapReduce performance with data compression: A study using wordcount job," 2017 14th International Conference on Electrical Engineering/Electronics, Computer, Telecommunications and Information Technology (ECTI-CON), 2017, pp. 564–567
32. E. Zagan and M. Danubianu, "Data Lake Approaches: A Survey," 2020 International Conference on Development and Application Systems (DAS), 2020, pp. 189–193.
33. R. Shree, T. Choudhury, S. C. Gupta and P. Kumar, "KAFKA: The Modern Platform for Data Management and Analysis in Big Data Domain," International Conference on Telecommunication and Networks (TEL-NET), 2017.
34. A. Sharma, A. Jain, P. Gupta and V. Chowdary, "Machine learning applications for precision agriculture: A comprehensive review," in IEEE Access, vol. 9, pp. 4843–4873, 2021.
35. R. N. Bashir, I. S. Bajwa and M. M. A. Shahid, "Internet of things and machine-learning-based leaching requirements estimation for saline soils," IEEE Internet of Things Journal, vol. 7, no. 5, pp. 4464–4472, May 2020.

Part II

---

*Data-Driven Techniques
and Security Issues in
Wireless Networks*

6 Data-Driven Techniques and Security Issues in Wireless Networks

Mamoon M. Saeed
Communication and Electronics Engineering Department, University of Modern Sciences, Yemen

Elmustafa Sayed Ali
Electrical and Electronics Department, Red Sea University, Sudan

Rashid A. Saeed
Computer Engineering Department, Taif University, Saudi Arabia

CONTENTS

DOI: 10.1201/9781003216971-8

6.1   INTRODUCTION

In many areas of wireless networking, data-driven approaches have become critical. They have evolved to become necessary in certain situations that require high confidentiality [1]. Cell phones have already surpassed landline/corded phones in terms of number, and smartphones will soon outnumber personal computers (PCs). For the past two decades, securing wireless networks has been an important research area, with no clear answer as to which security technique should be utilized to avoid unauthorized data access [2, 3]. Wireless security is the process of leveraging wireless networks, such as Wi-Fi networks, to prevent illegal access or damage to machines or data. Wi-Fi security, which includes wired network equivalent privacy (WEP) and wireless secure access, is the most often-used type of network security. WEP is a well-known security standard that provides insufficient protection and security [4]. Many portable computers have wireless cards pre-installed, which allow these devices to access wireless networks. However, wireless networks are susceptible to some security issues. Computer hackers have discovered that wireless networks are relatively easy to hack, even using wireless techniques to penetrate wired networks. As a result, organizations need to establish effective wireless security policies to protect against unauthorized access to critical resources and assets [5, 6].

In general, wireless networks are not secure, and the data transmitted through them can be easily hacked and modified. In wireless networks, security is more important, and some countermeasures are required to maintain confidentiality. If the data is sensitive and related to wireless networks of financial institutions, banks, or military networks, additional measures must be taken to ensure the confidentiality and security of information [6]. Moreover, when moving to an Internet of Things (IoT) network, and considering the multiple services provided by the IoT with great advantages, the security risks that come with Internet connectivity are a source of concern, as there are so many defined security threats on the Internet.

In recent years, many notable IoT attacks have been documented that have severely damaged equipment or systems connected to the Internet. There are concerns about security holes in devices that can allow devices to be modified by hackers, which is one of the most common problems related to IoT security [12]. Therefore, many studies have developed techniques and methodologies for network security and cybersecurity that have been beneficial in IoT systems. Furthermore, organizations' wireless systems are required to be examined and observed to assess their actual behavior using a data-driven strategy [7].

Because of the importance of data-driven approaches and security issues in wireless systems and networking, this chapter provides a comprehensive idea about the data-driven strategy with some brief security concepts in wireless networks. The chapter is organized as follows; Section 6.2 presents the motivation study and related works for data-driven techniques and security issues in wireless networks. Sections 6.3 and 6.4 discuss the data-driven concept in wireless networks and the open systems interconnection (OSI) model in network security. The data-driven security challenges, attacks, and threats in data-driven wireless technology are presented in Sections 6.5, 6.6, and 6.7. Moreover, the chapter also reviews some data-driven security solutions and discusses wireless access protocol (WAP) technologies in Sections 6.8 and 6.9. A data-driven defense approach toward secure IoT devices, the requirements for data-driven security and privacy in wireless body area networks WBANs and distributed data-driven security are discussed in Sections 6.10, 6.11, and 6.12, respectively. In addition, in Sections 6.13 and 6.14, the chapter provides moving toward the future of data-driven security and intelligence. Finally, a chapter summary is given in Section 6.15.

6.2   MOTIVATION STUDY AND RELATED WORKS

Recently, an increasing number of research studies in the wireless security area have relied on massive data sets to verify their premise, whereas other "early adopters" of wireless security have avoided discussing data-driven security. Because data-driven research has found applications in a variety of wireless network methods, it is necessary to investigate and improve wireless security to protect data-driven research. As a result, the fundamental contribution of this chapter is to provide a well-structured starting point for implementing data-driven concepts in wireless networks, to reduce the impact of attackers and hackers. There are many studies [7-10] that presented an anonymous and lightweight authenticated key agreement mechanism for smart-grid-based demand response management. Such studies discussed surviving known security threats while also allowing for privacy and mutual authentication. They used informal security analysis to evaluate the supplied protocol's security qualities [10]. Whereas in another study presented by S. A. Chaudhry *et al.*, a certificate-based device-to-device (D2D) access control method for Internet of Medical Things (IoMT) systems is suggested using elliptic curve encryption [11]. Formal and informal methods are used to prove the security of the proposed D2DAC-IoMT.

In study by S. A. Chaudhry *et al.* the authors developed a superior strategy and demonstrated its security. The suggested technique completes the authentication cycle with only a tiny increase in calculation cost while meeting all security and privacy criteria [12]. However, on the Internet of Drones (IoD) domain, researchers devised a general certificate-based access control mechanism to offer inter-drone and drone-to-ground station access control/authentication (GCACS-IoD) [13]. The GCACS-IoD provides anonymity and is provably secure against known attacks. GCACS-IoD improves security while maintaining computation and communication performance. On the other hand, the integration of wireless technologies into traditional medicine, including the diagnosis, monitoring, and treatment of sickness, is changing health care. The number of instruments available for remotely monitoring and detecting disease is growing. The capacity to handle medications and health devices remotely is becoming more common. The understanding of how genetics affects illness susceptibilities is growing thanks to smart and intelligent systems. These developments suggest that they are on the verge of a healthcare revolution [14].

Fog computing inherits cloud computing's privacy and security challenges, such as authentication and key management issues. To solve these challenges, Z. Ali et al. offered an enhanced approach to address these issues while maintaining its strengths. Simultaneously, an informal security study was carried out to ensure that the plan could withstand recognized threats [15]. The suggested approach is also compared to some state-of-the-art schemes using computation and communication costs as metrics. In contrast, M. Rana *et al.* claimed that the Kaul and Awasthi protocol is insecure because an attacker can readily discover the identity of a legitimate user sending data via the public channel [16]. Furthermore, an attacker can impersonate a valid user of the system and access the server's services by utilizing the identity of a legitimate user. As a result, their protocol is vulnerable to impersonation attacks, and their claim that it is secure is debunked. As a result, they have expanded on their work and presented an improved technique that ensures secure communication throughout the channel.

R. Majeed et al. introduced a revolutionary smart home concept that employs a support vector machine (SVM) for intelligent decision making and blockchain technology to ensure IoT device identification and authentication [17]. Emerging blockchain technology plays a critical role in the proposed home automation system by offering a reliable, secure, and decentralized way for identifying and authenticating IoT devices. Whereas M. Sohail *et al.* proposed a trust enhanced on-demand routing (TER) scheme that uses the Trust Walker (TW) algorithm for efficient trust evaluation and route discovery. The TW algorithm and the ad hoc on-demand distance vector (AODV) routing protocol are all part of the TER [18]. The simulations show that the proposed approach is accurate in terms of throughput, packet drop ratio (PDR), and end-to-end (E2E) delay.

6.3   THE DATA-DRIVEN CONCEPT IN WIRELESS NETWORKS

To be defined as data-driven, a technique must focus on data and large groups of data to solve a specific problem by better understanding the behavior of complex systems that cannot be easily modeled or simulated. The study relies on generalized extraction of knowledge from data to achieve a specific goal after conducting many experiments on real-world data and using data-science techniques to build statistical models that can be used to better understand the behavior of the system and extract new knowledge [28]. Recently, the use of data-based research has increased in various fields, especially in communications analysis applications in wireless networks and cloud computing applications, in addition to traffic analysis in mobile-cellular networks. Wireless network applications, due to their unpredictable nature, have become a place of interest for many researchers, especially in data science and its relationship to communication systems [30].

Data-driven techniques are used to comprehensively examine huge data in wireless sensor network applications and IoT networks. It achieves a step in the overall process of data knowledge discovery. The algorithm must be selected, implemented, and evaluated as possible and meaningful only after the problem is well defined and the data, including statistical characteristics, are well analyzed and understood [40]. In wireless networks, machine learning (ML) is used for perceptual processes, context awareness, and intelligence capabilities in various aspects of wireless communications. Learning techniques help improve network-level performance and service quality. Intelligent behavior can also be used by adapting to complex and dynamic changing environments (wireless) and adding automation to achieve concepts of self-healing and self-improvement [42].

In recent years, several different learning approaches have been used in many wireless network systems, such as routing access control, data collection, energy harvesting, cognitive radio, and IoT networks with their different applications. The ML model in wireless networks helps to form networks in real-time format with near-perfect performance, but sometimes there is a sharp decrease in the amount of data required to train ML-based models through the use of this hybrid, even when the available analytic model is imprecise. However, there are other advantages to the combined ML and analytical models. The reinforcement learning (RL) framework shown in Figure 6.1 helps extrapolate advantages to other ML techniques. Even in the case that some performance measures are difficult or not measurable directly from the network, analytical models with RL can calculate the relevant features of ML.

Several features of RL can be offered in federated learning (FL) frameworks to deal with sequential decision-making tasks, thus, federated reinforcement learning (FRL) can be considered a fusion of FL and RL under privacy protection [8]. FRL is an ML algorithmic framework that allows multiple parties to do ML while adhering to privacy, data security, and regulatory requirements. Model creation in FRL architecture is divided into two steps: model training and model inference. During training, parties can exchange information about the model but not the data itself, ensuring that data privacy is not jeopardized in any manner. The trained model might be owned and maintained by a single person or numerous parties. More data instances acquired from diverse parties contribute to updating the model during the model aggregation phase [9].

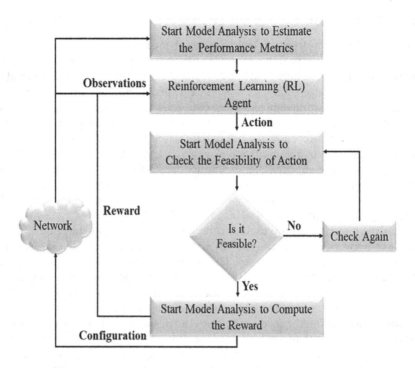

FIGURE 6.1   Reinforcement learning (RL) framework for wireless networks.

6.4   OSI MODEL IN DATA-DRIVEN SECURITY

Administrators require framework standards to implement numerous protocols to appropriately ensure the integrity of a network. To replace TCP/IP and meet this criterion, the open systems interconnection (OSI) model was established as a network reference model for studying data-driven flow between hardware and software in a seven-layer system. The seven levels of architecture and functions are shown in Table 6.1. The layer in the OSI model is vulnerable to a variety of attacks, which causes a network's performance to stall [7].

In the OSI model, each layer is responsible for supporting the layer above it and giving service to the layer below it, in addition to fulfilling particularly specialized responsibilities [8]. OSI layers are divided into two groups based on their functions: layers 1–4 are assigned to the lower layers of protocol stacks and media layers responsible for data-driven functions, and layers 5–7 are associated with application-level data and are considered the upper host layers of the system.

6.4.1   PHYSICAL LAYER VULNERABILITIES

Threats include physical theft of data and devices, physical damage or destruction of data and devices, unauthorized changes to the functional environment involving data communications, removable media, and adding or removing resources. Moreover,

TABLE 6.1

OSI Model, Seven Layers, and Architecture

Layer Type	Layer Number	OSI Model Layer	Data-Driven Unit Type	Function
Host Layers	7	Application	Data	• Application interface • Interpreting program requests, info requirements
	6	Presentation	Data	• Data compression • Data representation • Encryption
	5	Session	Data	• Inter-host communications
Medium Layers	4	Transport	Segments	• End-to-end connection • Sequence of packets
	3	Network	Packets/ Datagram	• Establish network connection • Individual packets transmission • Logical addressing IP
	2	Data Link	Bit/Frames	• Psychical addressing
	1	Physical	Bits	• Binary bit transmission • Physical network connection • Media access

threats are related to the disconnection of physical data links, undetectable data-driven interception, and other input logging. The processes of frequency selection, carrier frequency production, data-driven encryption, signal detection, and modulation are all handled by the physical layer [19]. The threats in WSNs are susceptible to jamming, much like any other radio-based media. Furthermore, nodes in WSNs may be put in insecure or hostile areas with simple physical access to an attacker. In this section, we will look at these two flaws in more detail.

a. **Jamming**

Jamming is a type of assault that affects a network's radio frequency [20]. A jammer source can be powerful enough to interrupt the entire network, or it might be weaker and simply disturb a segment of the network. Even with low-powered jamming sources, such as a small compromised subset of the network's sensor nodes, an attacker can disrupt the entire network if the jamming sources are distributed randomly throughout the network. Jamming defenses include frequency hopping and code spreading, both of which are spread-spectrum communication variations [5]. Frequency-hopping spread spectrum (FHSS) is a form of signal transmission that involves rapidly switching a carrier among many frequency channels while employing a pseudo-random sequence that both the transmitter and receiver are aware of. In addition, an attacker may instead jam a large portion of the frequency band. Another strategy for defending against jamming assaults is code spreading, which is widely employed in mobile networks. However, because this technique necessitates more design complexity and energy, it is not suitable for usage in WSNs.

b. **Tampering**

Another physical layer of assault is tampering [5]. With physical access to a node, an attacker can retrieve sensitive data such as cryptographic keys and other data. The attacker might even modify or replace the node to create a compromised node under his or her control. One mitigation against this attack is to tamper-proof the node's physical package [5]. The sensor nodes in WSNs are typically expected to be tamper-proofed due to the additional expense.

6.4.2 LINK LAYER VULNERABILITY

Media access control (MAC) address spoofing, described as one station assuming the identity of another and the virtual local area networks (VLAN) circumvention, is a station that may force direct communication with other stations, bypassing logical controls such as subnets and firewalls. Spanning-tree faults can occur accidentally or intentionally, causing the second-layer environments to broadcast packets in an unending loop. Protocols in layer two during wireless media scenarios may allow unauthorized organizations to connect to the network for free, or insufficient authentication and encryption may offer a sense of security [21]. Switches may be forced to flood traffic to all VLAN ports rather than selectively forwarding to the relevant ports, allowing any device connected to a VLAN to intercept data. Multiplexing of data streams, data frame detection, media access, and error correction are all handled by the data link layer [19]. In a communication network, it enables reliable point-to-point and point-to-multipoint connections. At the connection layer, there are attacks such as intentionally causing collisions, resource fatigue, and unfairness. This section delves into each of the three types of link-layer attacks [21].

a. **Collisions**

When two nodes try to communicate on the same frequency at the same time, a collision occurs [22]. When packets collide, a change in the data-driven component is likely to occur, resulting in a checksum mismatch at the receiving end. After that, the packet will be discarded as invalid. An attacker might strategically cause collisions in specific packets, such as acknowledgment (ACK) control messages. Certain MAC protocols may experience costly exponential backoff because of such collisions. Error-correcting codes [23] are a common defense against collisions. Low levels of collisions, such as those produced by environmental or probabilistic mistakes are best for most codes. These codes, on the other hand, add to the processing and transmission overhead. It is safe to suppose that an attacker will always be able to corrupt more than what can be repaired. While it is feasible to detect intentional collisions, there are yet no complete defenses against them.

b. **Exhaustion**

An attacker can also utilize repeated collisions to cause resource exhaustion [22]. A naive link-layer implementation, for example, would keep attempting to retransmit the malformed packets. The energy reserves of the transmitting node and those around it will swiftly decrease unless these fruitless retransmissions are found and prevented. One proposed approach is to employ rate restrictions in the MAC admission control so that the network can reject excessive requests and save energy waste from repeated broadcasts [22]. A second method is to use time-division multiplexing, in which each node is given a specific time slot in which to communicate [22]. This eliminates the requirement for frame-by-frame arbitration and allows a backoff algorithm to address the infinite postponement problem. It is, however, still vulnerable to collisions.

c. **Unfairness**

Unfairness is a type of denial of service (DoS) attack that is deemed weak [24]. Intermittently using the previously mentioned link-layer assaults, an attacker can generate unfairness in a network. An attacker can degrade service to obtain an advantage, such as forcing other nodes in a real-time MAC protocol to miss their transmission deadline. By decreasing the length of time an attacker has to capture the communication channel, tiny frames reduce the impact of such attacks. However, when an attacker tries to retransmit fast instead of arbitrarily waiting, this strategy generally loses efficiency and is vulnerable to more injustice [24].

6.4.3 NETWORK LAYER VULNERABILITIES

Routing spoofing is the dissemination of fictitious network topology. False source addressing on malicious packets is known as IP address spoofing. Identity and resource identification vulnerability relying on addresses to find resources and peers might be dangerous. Sensors networks and routing layers are usually developed according to energy efficiency approach, where energy is a major factor associated with sensors and data networks primarily. In addition, the ideal sensor network uses attribute-based addressing and is aware of its location. The following are examples of network and router layer attacks.

a. **Spoofed, Altered, or Replayed Routing Information**

Targeting the routing information itself while it is being transmitted between nodes is the most direct attack against a routing protocol in any network. To interrupt network traffic, an attacker can fake, change, or replay routing information [25, 26]. Routing loops are created, network traffic is attracted or repelled from certain nodes, source routes are extended and shortened, fake error messages are generated, the network is partitioned, and end-to-end latency is increased. A message authentication code is appended after the message as a countermeasure against spoofing and tampering. The receivers can check whether the messages have been faked or altered by adding a message authentication code to the message. Counters or time-stamps can be embedded in messages to protect against replayed data-driven [27].

b. **Selective Forwarding**

The assumption that all nodes in the network will appropriately convey received messages is a big one in multichip networks. An attacker could design rogue nodes that only forward certain messages while dropping others [26]. The black hole attack is a type of this attack in which a node drops all messages it receives. Using several paths to convey data-driven messages is one defense against selective forwarding attacks [26]. The second line of defenses is to detect the malicious node or presume it has failed and look for another route.

c. **Sinkhole**

An attacker uses forged routing information to make a compromised node appear more desirable to nearby nodes in a sinkhole attack [26]. As a result, adjacent nodes will choose the compromised node as the next node via which to route their data. Because all communication from a vast area of the network will pass through the adversary's node, this form of attack makes selective forwarding straightforward. The Sybil attack occurs when a single node presents the network with many identities [20, 26, and 28]. Fault-tolerant methods, distributed storage, and network-topology maintenance are among the protocols and techniques that are easily affected. To achieve a certain amount of redundancy, a distributed storage strategy might rely on three clones of the same data. If a compromised node impersonates two of the three nodes, the algorithms may incorrectly believe that redundancy has been achieved.

d. **Wormholes**

An attacker can repeat network messages using a wormhole, which is a low-latency link between two parts of the network [26]. This connection can be made by a single node relaying messages between two neighboring but otherwise unrelated nodes, or by two nodes in separate areas of the network talking with each other. The latter instance is similar to the sinkhole attack in that an offensive node close to the base station can establish a one-hop link via another network-attacking node to that base station. Hu et al. proposed packet leashes as a new and generic approach for detecting and guarding against wormhole assaults [29]. Geographic and temporal leashes were the two types of leashes introduced.

e. **HELLO Flood Attacks**

Several procedures that rely on HELLO packets create the erroneous assumption that receiving one signifies the transmitter is inside the radio transmission range and thus a neighbor. A high-powered transmitter could be used by an attacker to deceive a large number of nodes into trusting they are neighbors of the broadcasting node [26]. If the attacker fraudulently broadcasts a better path to the base station, all of these nodes will try to communicate with the attacking node, even though many of them are out of radio range in actuality.

f. **Acknowledgment Spoofing**
Acknowledgments are occasionally required by routing algorithms in sensor networks. To send false information to surrounding nodes, an attacker node can fake overheard packet acknowledgements for those nodes [26]. Claiming that a node is alive when it is dead is an example of misleading information.

6.4.4 Transport Layer Vulnerabilities

One of the transport layer vulnerabilities is the mismanagement of ambiguous, undefined or illegal situations that arise during the exchange of data by some applications. where, there can be a threat through transmission protocols that can fingerprinting the host information and use it to carry out attacks on the network. The ability to efficiently filter and qualify traffic is hampered by overburdening transport-layer methods, such as port numbers. Transmission systems can be spoofed and attacked using designed packets and educated guesses of values of flow and transmission, allowing communications to be disrupted or taken over. End-to-end connections must be managed by the transport layer [19, 30]. This subsection discusses two possible assaults in this layer: desynchronization and flooding.

a. **Desynchronization**
The breakdown of an established connection is referred to as desynchronization [24]. An assailant, for example, could send communications to an end host that are spoofed frequently, prompting that host to seek retransmission of frames that were missing. An attacker can degrade or even prevent end hosts from effectively exchanging data-driven messages if they are timed appropriately, leading them to waste energy attempting to recover from faults that never existed. Authentication of all packets sent between hosts [24] could be a feasible solution to this kind of assault. An attacker will be unable to send the faked messages to the destination hosts if the authentication technique is safe.

b. **Flooding**
When a protocol must keep state on both ends of a joining, it is prone to memory exhaustion due to flooding [24]. An attacker can keep requesting additional connections until each connection's resources are consumed or a limit is reached. Further reasonable requests will be disregarded in either situation. One solution to this issue is to make each connecting client solve a puzzle to establish their commitment to the connection [24]. The concept is that a connecting client will waste no resources by making connections that are not necessary. Given that an attacker is unlikely to have infinite resources, creating new connections quickly enough to trigger resource hunger on the serving node will be unfeasible. While there is some processing cost in these puzzles, it is preferable to an excessive amount of communication.

6.4.5 Session Layer Vulnerabilities

Authentication is generally a weak or non-existent measure, and it is possible to pass the clear text of session credentials such as user ID and password, allowing for unauthorized use spoofing and session identification hijacking. Information leakage due to unsuccessful authentication attempts can become vulnerable to violent attacks on access credentials and cause an unlimited number of failed sessions. The attacker attempts to extract the data transmitted between the user's device, the wireless access point, and the authentication server by using penetration-testing software to capture data pushed between these devices. Then the attacker could extract the key to find a way to forcibly disconnect the normal user and replace it with a legitimate file via a terminal to gain greater operational privileges over local networks [31]. The first attempt, which used a convolutional neural network (CNN) to obtain the relationship between the MAC address and the IP address, yielded the address resolution protocol (ARP) routing table.

6.4.6 PRESENTATION LAYER VULNERABILITIES

Unpredicted input can cause a program to crash or surrender control, allowing arbitrary instructions to run. The accidental or ill-advised use of externally supplied input in control settings may result in remote manipulation or information leakage. Cryptographic weaknesses could be used to get around privacy safeguards [32].

6.4.7 APPLICATION LAYER VULNERABILITIES

Unintentional use of program resources is made possible by open design concerns. Standard security safeguards are bypassed through backdoors and program design faults. Overly complicated application security controls are often overlooked or misunderstood and executed, and insufficient security controls compel an "all-or-nothing" strategy that results in either too much or too little access. Program logic faults can be exploited to crash programs or induce unwanted behavior, either accidentally or by design [33].

6.5 DATA-DRIVEN SECURITY CHALLENGES, ATTACKS, AND THREATS

Before several last years, the military was the primary client for wireless security products, particularly during the Cold War; network threats were not known to the general public until wireless equipment prices fell around the year 2000; however, today, everyone, including businesses and the military, is well aware of network security.

6.5.1 WORM

A worm and a virus have numerous traits in common. The most crucial feature is that worms are also self-replicating; however, worm self-replication differs from a virus in two ways. To begin with, worms are self-contained [34] and do not require any additional executable programs. Second, worms move across networks from machine to machine. The first worms, like viruses, were human made.

6.5.2 DISTRIBUTED DENIAL OF SERVICE (DDoS)

Server failures and financial losses can result from distributed denial of service (DDoS) attacks, which put IT personnel under a lot of pressure to get resources back online. The correct detection and prevention techniques can help halt a DDoS attack before it becomes powerful enough to bring down a company's network [35]. Websites and online services are the targets of DDoS attacks where the goal is to overwhelm them with traffic that exceeds the server or network's capacity. The intention is to make the website or service inaccessible. Incoming messages, connection requests, and bogus packets might all be part of the stream in the attack.

6.5.3 SQL INJECTION

SQL injection, often known as SQLi, is a frequent method used by attackers to change and access database information that would otherwise be hidden or unavailable to website visitors [36]. This is performed by leveraging application vulnerabilities to insert malicious SQL code into SQL queries to extract, edit, or delete data-driven information that the logged-in user does not have permission to access.

6.5.4 SPOOFING

A spoofing attack is a circumstance in which a person or computer effectively identifies as another by misrepresenting data-driven messages to achieve an unfair advantage in the context of information security, particularly network security. Many TCP/IP protocols lack means for authenticating the message's source or destination, making them vulnerable to spoofing attacks if applications do not take extra efforts to authenticate the identity of the sending or receiving host. Man-in-the-middle attacks against hosts on a computer network can be leveraged using IP spoofing and ARP spoofing in particular. Spoofing attacks that use TCP/IP suite protocols can be countered by using firewalls with deep packet inspection capabilities or by taking steps to authenticate the sender or recipient of a message's identity [37].

6.5.5 EAVESDROPPING

Eavesdropping, also known as a sniffing or a spying attack, is when a computer, smartphone, or other connected device steals information while it is transferred over a network. The attack uses access gained to data-driven information as it is transferred or received by the user through unsecured network communications [38]. The term "eavesdropping" is deceptively gentle. Typically, the attackers are looking for sensitive financial and corporate data-driven information that can be sold for illicit purposes. "Spouse-ware," which allows people to eavesdrop on their loved ones by tracking their smartphone usage, is also a big business [39]. Because the network transmissions appear to be regular, an eavesdropping assault can be difficult to detect. An eavesdropping attack involves a weakened link between a server and a client, which the attacker can exploit to reroute network traffic to be successful.

6.5.6 MALWARE

Malware can be classified into many types based on how it operates. Despite its name, antivirus software cannot detect all of these varieties of malware [40]. Self-replicating malware actively tries to spread by making new copies of itself. Malware can also spread passively, such as when a user copies it by accident, but this is not self-replication [41]. The population growth of malware due to self-replication indicates the total change in the number of virus instances. Malware that does not self-replicate will always have a population growth of zero, however, malware that does self-replicate may have a population growth of zero [42].

6.5.7 BRUTE FORCE

A brute-force assault in cryptography entails an attacker submitting a large number of passphrases or passwords in the hopes of accurately attempting to guess a combo. The attacker goes through all conceivable passwords and passphrases to find the proper one. Alternatively, the attacker can use a key derivation function to guess the key, which is normally generated from the password. This is referred to as a thorough key search. A brute-force assault is a cryptographic technique that can theoretically be used to decrypt any type of encrypted material except for data-driven material, which is encrypted in a way that is theoretically secure [42]. When other flaws in an encryption scheme that could make the job easier (if any exist) are unavailable, this type of attack may be performed. When checking all short passwords, this method is very fast. However, for longer passwords, alternative approaches such as the dictionary assault are employed because a brute-force search is too time-consuming for longer passwords and keys because the larger number of possible values makes them exponentially more difficult to crack [42]. Brute-force assaults can be rendered less successful by obfuscating the data-driven information to be encoded, making it more difficult for an attacker to recognize when the code has been broken, or by requiring the attacker to rigorously evaluate each guess. The amount of time it would take an attacker to successfully conduct a brute-force attack against an encryption system is one of the metrics of its strength [43]. Table 6.2 summarizes the attacks categories and their impact to the OSI Model.

6.6 REVERSE ENGINEERING CHALLENGES IN DATA-DRIVEN SECURITY

Reverse engineering is the process of creating something from nothing. It is also known as backward engineering or back engineering, which is a technique or approach for attempting to know what a device, process, system, or piece of software does—a goal using deductive reasoning without providing much insight into how it accomplishes this. Chemical engineering, systems biology, computer engineering, mechanical engineering, design, electronic engineering, and software engineering are all domains [44] where reverse engineering can be used [45]. Reverse engineering is used for a variety of purposes in many fields. The beginnings of reverse engineering can be traced back to the investigation of commercial or military hardware purposes [46]. The reverse

TABLE 6.2

Summary of Attacks Categories and Impact on OSI Layers

Attack	Capabilities and Effects	Layers Impacted	References
Worm	Network traversal: Disrupting disparate system components	2	[48–51]
DDoS	Network operation disruption: Effect on system resources	2	[50–53]
SQL Injection	Invalid data insertion and reduces system integrity	4, 5, and 6	[50, 51]
Spoofing	System infiltration effect	1, 2, 3, and 7	[50, 51]
Eavesdropping	Unauthorized interaction system	1, 2, 3, and 7	[50, 51, 53]
Jamming	Communication channel saturation and improper operations	1, 2, and 3	[50, 51, 54]
Malware	Infiltration of malicious codes	1, 3, 4, 6, and 7	[50, 53]
Brute Force	Repetitive attempts to infiltrate the system	1, 3, 6, and 7	[50]
Reverse Engineering	Attempts to extract the weakness in network actions and design	1, 3, and 7	[50, 51, 53–55]

engineering technique, on the other hand, is not worried about making a replica or altering the artifact in any manner. It's just a process of deducing design elements from items created by people who had little or no prior awareness of the processes utilized to create them. The purpose of the reverse engineering process might often be as simple as re-documenting bequest schemes [47]. Even if the competitor's product is reverse-engineered, the purpose may be to undertake competition analysis rather than to imitate it. Despite some narrowly tailored legislation in the United States and European Union, reverse engineering can be used to create interoperable products, and the legality of using specific reverse engineering techniques for that purpose has been hotly debated in courts all over the world for more than two decades [45].

Program reverse engineering can help with software maintenance and improvement by helping to understand the underlying source code. Graphical representations of the code can provide other views of the source code and essential information can be retrieved to aid with software development decisions that can aid in the detection and correction of a flaw or vulnerability in the software. Design information and advancements are typically lost as the software matures; however, reverse engineering is generally used to extract this information. The technique can also aid in reducing the time it takes to comprehend source code [44]. With stronger code detectors, reverse engineering can also help discover and delete harmful code written to software. Reversing source code is a type of code that can be used to discover alternative uses for the code, such as discovering illicit replication of the code in places where it was not meant to be used or showing how a competitor's product was constructed [45]. This method is widely used to "crack" software and media to get rid of the copyright protection or to generate a potentially improved copy or even a copycat, which is normally the goal of a rival or a hacker [46].

Malware developers frequently employ reverse engineering techniques to identify operating system flaws to create a computer virus that can take advantage of those flaws. In cryptanalysis, reverse engineering is also used to uncover flaws in symmetric-key algorithms, public-key cryptography, and substitution ciphers [44]. In an already complex Internet security landscape, IoT technologies provide new concerns. The sheer increase in the number of linked IoT devices poses a substantial difficulty in and of itself, as previously highlighted. Furthermore, the proliferation of IoT systems is regularly bringing connectivity to new classes of devices, each of which may create distinct issues that must be addressed. Some are evident, while others are less so [56]. As new products come online that were previously exclusively available offline, IoT is pushing the Internet's edge farther into our daily lives. Because of the rapid rate at which they are being introduced, testing and proper security design may take a back seat to the desire to get new goods into the market ahead of the competitors. As if that weren't bad enough, unlike their predecessors, IoT devices are typically located outside of the relatively secure confines of secure data centers, putting them within easy reach of those seeking to exploit any vulnerabilities that may exist within the IoT system through physical access [57]. Beyond these issues, the fact that IoT devices are frequently found in our homes, offices, and public spaces means that if they are hacked, they can be used to spy on people who are close to the gadgets.

The vulnerability of IoT systems varies dramatically depending on where they fall in the seven-layer reference architecture. Direct gadgets are vulnerable to a variety of attacks due to their physical accessibility to attackers. An attacker can physically obtain access to an edge device, tamper with it, and possibly reverse engineer it to gain access to the entire system. Because the Internet is utilized as a communications medium, the communications layer is also an obvious attack surface. The Internet is known to be vulnerable to a variety of protocol assaults, ranging from well-known attacks to unknown zero-day attacks [58]. Attack tactics aimed at infiltrating or disrupting edge computing, data aggregation, or data abstraction are less common, although these layers are nonetheless vulnerable to software vulnerabilities lying in third-party programs and devices. SQL injections and other forms of attacks that were formerly used to enter data stores can still be employed against poorly built software. Applications are subject to the same issues and must be properly vetted to prevent attacks that take advantage of poorly designed software, firmware, or hardware [59].

6.7 GENERAL ATTACKS/THREATS IN DATA-DRIVEN WIRELESS NETWORKS

An assault is a physical action taken by an intruder with the intent of compromising information in a company. Wireless local area networks (WLANs), unlike wired networks, communicate through radio frequency or infrared transmission technologies, rendering them vulnerable to assault. These assaults try to compromise information confidentiality and integrity, as well as network availability. The two types of attacks are classified as active attacks and passive attacks. Active attacks are those in which the attacker not only gains access to the information on the network but also actively modifies it or even creates bogus information on the network. Any organization suffers a significant loss as a result of this type of malicious behavior [60]. Passive attacks are those in which the attacker seeks to get information from the network while it is being transmitted or received. Because the attacker does not alter the contents of the file, these attacks are frequently difficult to detect [61]. Passive attacks include eavesdropping and traffic analysis. Hijacking, rogue access point, the man in the middle (MITM), DoS, and reply attacks are current WLAN security attacks.

The confidentiality, integrity, and availability (CIA) triad holds that to establish complete information security, all three of those principles are required to some extent, otherwise the network will be susceptible to attack. Access control and authentication are two more principles that are involved. Data confidentiality refers to the prevention of data disclosure, whether intentionally or unintentionally. Data integrity refers to the ability to prevent data from being tampered with, whether intentionally or unintentionally. Availability refers to the ability to regulate the delivery of system resources to authorized users/systems/processes on demand. Authentication is the method by which a system confirms the identity of a user who wants to use it [62]. Access control is the control of access to resources by a legitimate user. Various sorts of threats/attacks in a WLAN are explored here; access control and authentication definitions are based on the CIA triad. These attack kinds can be either active or passive, as described previously.

6.7.1 CONFIDENTIALITY ATTACKS

Intruders seek to intercept highly confidential or sensitive information that is delivered across a wireless network using the 802.11 or higher layer protocols, encrypted, or in clear text. Eavesdropping, MITM attacks, traffic analysis, and other passive attacks are examples. Access Point (AP) phishing, WEP key cracking, and Evil Twin AP are examples of active attack categories [60]. Most hackers do traffic analysis, also known as foot printing, as the first step before launching additional attacks. Using this technique, the attacker identifies the communication load, the number of packets being communicated and received, the size of the packets being communicated and received, and the source and destination of the packets being transmitted and received. As a result, the traffic analysis attack [61] has gained access to the entire network activity. To detect the signal range, the attacker employs a wireless card that can be set to promiscuous mode, as well as specific types of antennas, such as a Yagi antenna, and the global positioning mode (GPS), to carry out this assault.

Furthermore, a variety of freely available software, such as Nets tumbler, Kismet, and others, can be employed. Through traffic analysis, the invaders receive three types of information. They begin by determining whether there is any network activity. Second, he or she counts how many access points there are in the area and where they are located. If the AP's broadcast service set identifier (SSID) is not disabled, the SSID is broadcast to allow wireless nodes to join the network throughout the wireless network. A passive sniffer like Kismet may get all network information; any AP's name, location, and channel are all included, even if it was turned off. Finally, the type of protocol being utilized in the transfer, as well as the number, size, and type of packets being transmitted are all types of information that an attacker can discover from traffic analysis. Take, for example, an examination of transmission control protocol (TCP's) three-way handshake information [61].

a. **Eavesdropping**

An eavesdropping attack allows an attacker to monitor network traffic and read the contents of messages sent over it. The attacker passively monitors the wireless session and payload. The attacker can decrypt the message later if it is encrypted. The attacker can collect data on the packets, their source, destination, size, number, and transmission time are all included. More crucially, several directional antennas are available on the market that can identify 802.11 networks broadcasts from miles away in the correct conditions. This is an attack that physical security measures alone will not be able to prevent. Furthermore, this attack can be carried out distant from any organization's premises [61, 62].

b. **Man-in-the-Middle Attack**

A MITM attack can be used to access or modify confidential information in a session, therefore jeopardizing the data's confidentiality and integrity. This technique also compromises the confidentiality of indirect data. However, security solutions such as an IPsec or virtual private network (VPN), which only guard in opposition to direct data-driven privacy assaults, could be used by an organization. This real-time assault takes place while the target machine is in use. This attack can be carried out in a variety of ways. In step one, for example, the attacker disconnects the target's client session from the access point and compels them to reconnect. The target client tries to reconnect with the access point in step two, but it can only do so through the attacker's PC, which is acting as the access point. The attacker connects to the access point and logs in as the target client in the meantime. If there is already an encrypted tunnel, the attacker constructs two additional, one between the attacker and the target client and the other between the attacker and the access point. In this form of attack, the attacker poses as an AP and a legitimate user of the AP to the target client [61].

6.7.2 Access Control Attacks

This exploit tries to obtain unauthorized access to a network by evading the firewall and filters. The most prevalent types of attacks in this category include illegal access, war driving, rogue access points, and MAC address spoofing.

a. **Illegal Access**

In this case, the attacker is attempting to acquire access to the entire network rather than a specific user. The attacker may be able to gain access to certain services or privileges to which he or she does not have the authority to access. Furthermore, some WLAN architecture provides the attacker access not only to the wireless network but also to the network's cable component. This can be accomplished with the use of MAC spoofing, war driving, or rogue access points. This technique allows the attacker to launch a more harmful attack like a MITM attack [61].

b. **Rogue Access Point**
   An attacker uses an unprotected AP in public venues like airports, shared office areas, or outside an organization's building to intercept messages from genuine wireless clients who think it's a genuine authenticator. Because of this technique, the attacker could fool the real client by changing the valid client's SSID to the same as the target company's. Furthermore, the attacker creates a false access point using an unused wireless channel. It is simple to dupe naïve users to connect to a bogus access point. As a result, a user's credentials could be easily stolen [63, 64].

6.7.3 INTEGRITY ATTACKS

An integrity attack modifies data-driven information as it is being transmitted. The intruder tries to change, add, or delete data-driven or management frames to the network, i.e. forged control packets, to fool the receiver or to facilitate another type of attack [65]. The most common example of this form of assault is a DoS attack. Other types include session hijacking, replay assaults, 802.11-frame injection, 802.11 data replay, and 802.11 data erasure.

a. **Session Hijacking**
   When an attacker steals a legitimate user's authorized and authenticated network session, this is known as session hijacking. The authorized user believes that the session loss is due to a WLAN failure. As a result, he or she is completely ignorant that the attacker has taken over the session. This attack occurs in real-time, and the attacker has complete control of the session and can maintain it for as long as he or she wishes [61]. The attacker must complete two steps to properly execute the session hijacking attack. In the beginning, the attacker poses as a legitimate WLAN target. Successful eavesdropping on the target communication is required to obtain the necessary information. Second, to keep the legitimate target out of the session, the attacker floods the wireless node with a series of spoofed dissociate packets [61].

b. **Availability Attacks**
   This attack inhibits or disables genuine customers by denying them access to the network's required data. The most frequent sort of availability assault is a DoS attack, which targets a specific area of the network and renders it unavailable. In DoS, an attacker floods a legitimate client with false packets, incorrect messages and MAC addresses, or duplicate IP to impede or prohibit normal network connection [66]. Another kind of attack is RF jamming. An 802.11 network uses the unlicensed 5 GHz and 2.4 GHz frequency bands to communicate. In this form of assault, the attacker uses a powerful radio signal to jam the WLAN frequency, rendering access points ineffective [61]. As a result, legitimate users are unable to connect to the WLAN.

6.7.4 VALIDATION ATTACK

A validation attack occurs when an attacker obtains the identities and credentials of legitimate users to acquire access to a private or public WLAN, as well as services.

The attacker poses as a legitimate user once they have obtained the required information. As a result, you'll have access to all of the WLAN's authorized rights [60, 66]. There are many more attacks involving 802.11 technologies than those listed previously, and discussing them all is beyond the scope of this chapter. A WLAN, for example, is subject to upper-layer attacks. Wired or wireless networks may be used to send Trojan downloaders, fishing messages, and bulk mailing worms. On wireless devices, attackers can poison ARP and DNS caches. There are also additional types of attacks that try to take advantage of the wireless encryption standard [60].

6.8 DATA-DRIVEN SECURITY SOLUTIONS IN WIRELESS NETWORKS

There are several solutions for network security, some of them are reviewed by S. Yi et al. such as security-aware ad hoc routing (SAR), a new routing technique that incorporates security attributes as parameters into ad hoc route discovery, allowing the use of security as a negotiable metric to improve the significance of the routes exposed by ad hoc routing protocols [67]. According to what was presented by L. Zhou et al., they used replication and new cryptographic approaches, such as threshold cryptography, to create a highly secure and accessible key management service [68]. Y.C. Hu et al., proposed a generic mechanism for detecting and guarding against wormhole assaults called packet leashes as well as the TIK protocol, which implements leashes [69]. As recommended by J. T. Isaac et al., a simple option to safeguard vehicular ad-hoc networks is to use cryptographic techniques and procedures that are currently widely used to defend against common computer network threats [70]. J. Sen propose cryptography as a critical and dominating security instrument that provides authentication, confidentiality, integrity, and non-repudiation in an efficient and cost-effective manner [71]. Despite the fact that cryptographic primitives are thought to be secure, it is possible that there are attacks on them that have yet to be discovered. For example, in collision attacks on hash functions, it is difficult to secure SHA-1, pseudorandom number assaults, digital signature attacks, and hash collision attacks.

In study presented by L. Hu et al., the authors demonstrated a methodology for secure wireless network aggregation that is resistant to both intruder devices and single device key compromises [72]. Their protocol was created to function within the constraints of low-cost sensor device processing, memory, and power consumption, but also take advantage of wireless networking features as well as power irregularity across units. According to J. Zhu et al. mobile users only execute symmetric encryption and decryption; a new and efficient wireless authentication mechanism based on the hash function and smart cards was presented [73]. Only one round of message exchange is required between the mobile user and the visited network, as well as one round of message exchange between the visited network and the mobile user, according to their protocol. Centralized server-based authentication solutions, such as remote authentication dial-in, can be used when firewalls or VPN gateways are used in the security of WLANs, according to Y. M. Erten The remote authentication dial-in user service (RADIUS) server is unique in that it uses the location information in conjunction with user privileges in access control, and it chooses to determine the client's location from IP subnet information, other research that used GPS technology for a similar goal found this to be substantially more difficult [74].

W. Alliance presented the Wi-Fi-protected access addresses all known vulnerabilities in Wi-Fi network security and vastly improves data-driven security and access control on current and future Wi-Fi WLANs. In addition, it provides an instant, strong, standards-based, interoperable security solution that addresses all known errors in the original WEP-based security [75]. In study by P. Guo et al., variable threshold-value authentication (VTA) architecture was proposed as a novel design prototype in the direction of lightweight and tolerant authentication for service-oriented wireless mesh networks (WMNs), in which VTA's intrusion-tolerant ability was guaranteed to design a series of node-stimulated mechanisms to keep the threshold values t and n of the system private key unchanged [76]. Based on the aforementioned vulnerabilities and threats, it is critical to ensure that the wireless network is secure, whether it is for a home user or a business network. However, there has yet to be developed a true security solution that is now available. However, the actions below can be used as a guide to avoiding the majority of known vulnerabilities and some typical threats.

The security of a WLAN should be considered throughout the development lifecycle, from initial design and deployment to implementation, maintenance, and monitoring. The administrator must ensure that the WLAN client devices and access points in the enterprise have always complied with security standards and have followed normal security configurations. Moreover, to analyze the overall security of the WLAN, the company should conduct continuous attack and vulnerability monitoring as well as undertake periodic technical security assessments [77]. WLANs are protected from the biggest threats by using robust encryption standards. The ideal practice is to use WPA/WPA2 instead of WEP for Wi-Fi-protected access. Furthermore, rather than using WPA, because WPA-TKIP relies on the WEP encryption mechanism for backward compatibility, it is suggested to use the WPA2, AES-CCMP protocol [24]. However, when using WPA2-PSK, users must choose passwords that are strong, long, and difficult to guess. Larger businesses might consider using a certificate-based authentication technique like RADIUS, which allows users to access their controlled credentials while keeping the network safe from sharing [78]. Hackers are well versed in all manufacturers' passwords, usernames, and default SSIDs. As a result, altering the default SSID is an important step in protecting a home or business network. More significantly, when choosing a name, a user should aim to come up with something unique that does not expose any personal information about the owner, such as a house number, street name, or business name This could allow the hacker to pinpoint the network's actual location [79]. The access point is effectively hidden by disabling SSID. This means that to access the WLAN, the network name and password must be configured manually by the user. This is only a rudimentary defense, as anyone with access to sniffer software can uncover the concealed network name. Where WEP is the sole option available, however, additional router security can be controlled [24]. A VPN can provide end-to-end security for wireless devices connected to an open network, such as those provided by hotels, Starbucks, McDonald's, and other locations. Furthermore, rather than trusting in the security of the business partner, large businesses can benefit from using a VPN to protect data-driven transmission to a home or business collaborate WLAN. A VPN can also be beneficial for securing traffic sent by devices that often move between wireless and wired networks, such as smartphones [80–81].

Another technology is VLAN, which can be used to implement a security policy in a business wireless network. VLANs work by assigning distinct workgroups to different LAN frames. Within the corporate network, such tags determine where incoming frames can and cannot go. For example, if a company allows guests and consultants access, all traffic from that wireless LAN will be tagged, which will limit access to the public Internet while keeping it away from business data-driven services [80]. Network access control (NAC) is another authentication mechanism that can be used in conjunction with 802.1x and VLANs to provide an extra layer of security. Rather than filtering traffic based on IP addresses and port numbers, NAC restricts user access to network resources based on the sender's confirmed user identification, the status of the user's device, and the set policy. Network devices like ethernet switches, wireless access points, routers, and firewalls can still control access with NAC, but they are merely enforcing the NAC's decisions. NAC can be implemented in a variety of ways, including permitting or disallowing the use of a specified SSID or using 802.1x to route wireless clients to VLANs or specific subnets [80]. For detecting intrusions and informing system administrators, a wireless intrusion detection and prevention system might be a useful tool. Passive sniffing on the network is impossible to block with a standard firewall. Wireless intrusion detection system (WIDS)/ wireless intrusion protection system (WIPS) can be utilized as a watchdog, detecting and blocking new threats and malicious conduct. A VPN paired with WIDS/WIPS can provide an efficient security solution by continuously monitoring the network for anomalies. This offers an additional layer of data-driven privacy protection [80].

6.9 WAP TECHNOLOGIES IN WIRELESS TECHNOLOGY

The wireless application protocol (WAP) is a global, open standard developed by the WAP Forum to provide mobile users of wireless phones and other wireless terminals such as pagers and personal digital assistants (PDAs) with access to telephony and information services such as the Internet and the Web. WAP works with all kinds of wireless networks TDMA, CDMA, and GSM. WAP is based on current Internet protocols as much as feasible, including IP, XML, HTML, and HTTP. It also comes with security features. Version 2.0 of the WAP specification is currently available as of this writing [82]. The severe constraints of the gadgets and the networks that connect them have a considerable impact on the adoption of data-capable terminals and mobile phones. The gadgets' processors, memory, and battery life are all constrained. In addition, the user interface is restricted, and the displays are modest. When compared to wired connections, wireless networks have lower bandwidth, higher latency, and less consistent availability and reliability. Furthermore, all these characteristics differ significantly from one terminal device to the next, as well as from one network to the next. Finally, users of mobile and wireless information systems have distinct requirements and expectations than users of other systems of information. Mobile terminals, for example, must be exceedingly simple to operate—far simpler than PCs and workstations. WAP was created to address these issues.

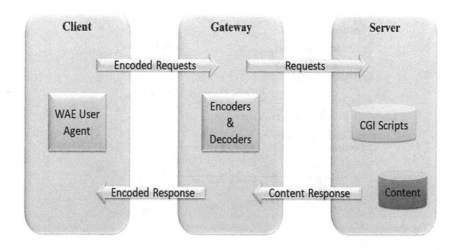

FIGURE 6.2   WAP programming model.

6.9.1   WAP OPERATIONAL CONCEPT

The client, the gateway, and the original server are the three components of the WAP programming model (see Figure 6.2). Between the gateway and the original server, HTTP is used to transport data. The gateway acts like a proxy server for the wireless domain. Its processor(s) perform functions that free up the limited capabilities of hand-held, mobile, and wireless terminals. For example, the gateway provides DNS services, converts between the WAP protocol stack and the web protocol stack (HTTP and TCP/IP), encodes information from the Web into a more compact form that reduces wireless communication, and decodes the compacted form back into standard Web communication conventions. The gateway also saves frequently requested information. [82].

The basic components of a WAP environment are depicted in Figure 6.3. WAP allows a mobile user to browse Web material on a normal Web server. The Web server serves content in the form of HTML-coded pages that are sent over the Internet utilizing the HTTP/TCP/IP protocol stack. The HTML content must pass through an HTML filter, which can be housed in a separate physical module or the same physical module as the WAP proxy. The HTML content is converted into wireless markup language (WML) content by the filter. If the filter is independent of the proxy, the WML is delivered to the proxy via HTTP/TCP/IP. The proxy compresses the WML and sends it to the mobile user using the WAP protocol stack via a wireless network. If the Web server can generate WML content directly, the WML is supplied to the proxy via HTTP/TCP/IP, WAP protocols are used to transform WML to binary WML, and data-driven information is subsequently sent to the mobile node. The architecture of the WAP was created to address two major restrictions of access to the Internet via wireless. These restrictions are, low rates in wireless digital networks, and the limitations of mobile node's related to restricted input capability,

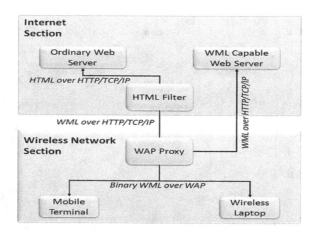

FIGURE 6.3   WAP infrastructure.

small screen size. Despite the development of 3G wireless networks that provide high-speed data-driven responses, input and display capabilities of small hand-held mobile nodes are still limited. As a result, WAP or a comparable capability will always be needed [82].

6.9.2   WAP ARCHITECTURE

In a WAP client, the entire stack architecture is implemented from the WAP architecture (see Figure 6.4). A five-layer model is presented here. Through a set of well-defined interfaces, each layer provides other services and applications with a set of functions and/or services. Other services and applications, as well as the layers above, can access each tier of the architecture. Several protocols may be used to deliver many of the services in the stack. The hypermedia transfer service, for example, could be provided via HTTP or wireless session protocol (WSP). Collections of services that are available by many tiers are common to all five layers. Security services and service discovery are the two types of common services [82].

WAP offers means for ensuring confidentiality, integrity, authentication, and nonrepudiation. The following are some of the security services available [82].

a. **Libraries for Cryptography:** This library at the application framework level provides data signing services for nonrepudiation and integrity.

b. **Authentication:** WAP enables a variety of client and server authentication options. At the session services layer, HTTP client authentication (RFC2617) can be used to authenticate clients to proxies and application servers. At the transport services layer, wireless transport layer security (WTLS) and transport layer security (TLS) handshakes can be used to authenticate clients and servers.

c. **WAP Identity Module (WIM):** WIM is responsible for storing and processing the information required for user identification and authentication.

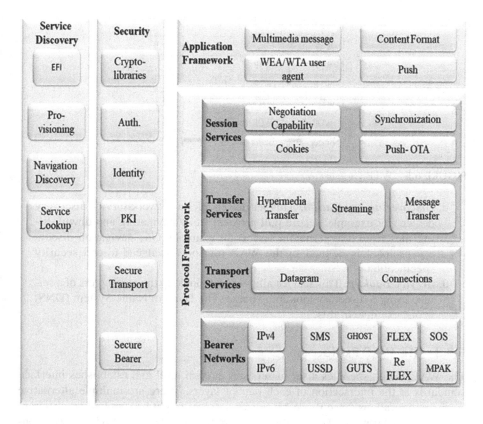

FIGURE 6.4 WAP architecture.

    d. *Public Key Infrastructure (PKI):* A set of security services that make public-key cryptography and certificates easier to use and administer.

    e. *Secure Datagram and Connection Transport:* The protocols of the transport services layer define secure datagram and connection transport. TLS is for secure transfer over connections, while WTLS is for safe transport over datagrams (i.e., TCP) [82].

    f. *Bearer Security:* Some bearer networks offer security at the bearer level. IPsec, for example, ensures IP network security at the bearer level (particularly in the context of IPv6).

The WAP client and the web server can use a set of service discovery services to determine what services and capabilities are available. The following are some instances of service discovery services [71].

    a. *External Functionality Interface (EFI):* The EFI allows apps to learn about the device's external functionalities and services.

    b. *Provisioning:* This service enables a device to be pre-configured based on the criteria required to use services for networks.

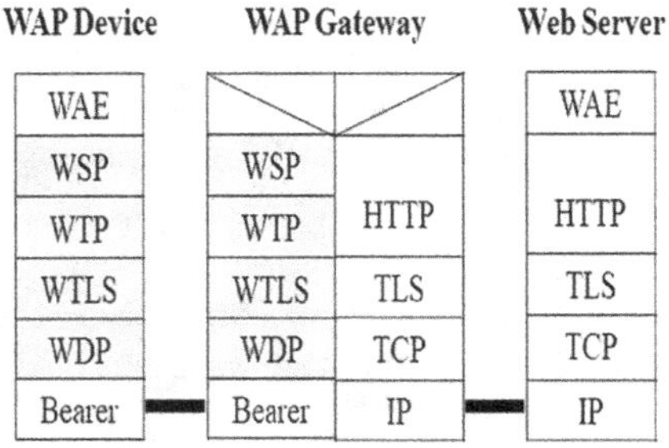

WAP Device   WAP Gateway   Web Server

WAE			WAE
WSP	WSP		
WTP	WTP	HTTP	HTTP
WTLS	WTLS	TLS	TLS
WDP	WDP	TCP	TCP
Bearer	Bearer	IP	IP

FIGURE 6.5   WTP 1.x gateway.

    c. ***Discovering Navigation:*** This service enables a network service discovery device (for example, secure pull proxies) while navigating and downloading files from a hypermedia server, for example. One navigation discovery protocol is defined in the WAP transport-level end-to-end security specification.

    d. ***Service Lookup:*** This service allows you to look up the parameters of a service using a directory lookup by name. The Domain Name System (DNS) is one example of this.

6.9.3   THE WAP PROTOCOLS

The WAP design specifies a set of services at each level and establishes interface standards at the intersection of each pair of layers. There are multiple alternative stack configurations since the WAP stack's services might be supplied via different protocols depending on the conditions. Figure 6.5 shows how a WAP gateway connects a WAP client device to a Web server in a typical protocol stack arrangement. In devices that implement Version 1 of the WAP specification, this arrangement is common, however, if the bearer network does not support TCP/IP, it can also be seen in devices that implement Version 2 (WAP2)[82].

6.9.3.1   Protocols for Wireless Session

The WSP provides a user interface for two different session services to applications. The session-oriented connection service runs on top of WTP, while the connectionless session service runs on top of the insecure WDP transport protocol. WSP is essentially HTTP with a few extensions and tweaks to make it work better over wireless channels. Low data rates and the risk of losing a connection due to poor coverage or cell overload are the main issues addressed. WSP is a protocol that focuses on transactions that use the request and reply concepts. A body that may or may not contain WML, WML script, or pictures, or a header that includes information about the material in the body as well as the transaction make up each WSP protocol data unit (PDU). Take, for example, a server push operation defined by WSP. In this scenario, the server sends unrequested content to a client device, which could be used to send out information or provide services that are personalized to each client's device, such as news headlines or financial prices [82].

6.9.3.2 Protocols for Wireless Session Transaction

For activities like surfing and e-commerce transactions, WTP facilitates transaction management by transferring requests and responses between a user agent, such as a WAP browser, and an application server. WTP provides a dependable transport service while eliminating much of the overhead associated with TCP, resulting in a lightweight protocol suitable for use in "thin" clients like mobile nodes and across low-bandwidth wireless networks [82]. WTP is equipped with the following features with three types of transaction services. First, optional user-to-user reliability, in which each received communication is confirmed by the WTP user. Out-of-band data-driven on acknowledgments is optional. To reduce the number of messages transmitted, PDU concatenation and delayed acknowledgment are used. WTP focuses on transactions rather than connections. WTP is a dependable connectionless service that does not require explicit connection establishment or deconstruction. WTP has three transaction types that WSP or another higher-layer protocol may use as reviewed in the following.

a. *Class 0*—The invoke message is unreliable and there is no result message: This class provides an unreliable datagram service that can be used to perform an unreliable push action. WTP (the initiator, or client) encapsulates data-driven messages from a WTP user in an invoke PDU and sends it to the target WTP (the server, or responder) without acknowledgment. The data is delivered to the target WTP user by the responder WTP.

b. *Class 1*—A reliable invoke message with no result message: This class offers a dependable datagram service that can be utilized for a dependable push operation. An invoke PDU encapsulates data from the initiator and sends it to the responder. The responder sends on the initiator side, an ACK PDU is sent to the WTP entity, which confirms the transaction with the source WTP user and acknowledges receipt of the data by delivering the data to the target WTP user. The responder WTP saves state information after the ACK is sent so that the ACK can be resent if it is lost or if the initiator retransmits the invoke PDU.

c. *Class 2*—An unreliable message to be invoked with a single, dependable outcome message: This provides a request/response transaction service and allows numerous transactions to be executed in a single WSP session. The data from the initiator is packaged in an invoke PDU and sent to the responder, who passes it on to the target WTP user. Response data is prepared by the target WTP user and passed on to the local WTP entity. These data are returned as a result of PDU by the responder WTP entity. If the answer data generation takes longer than the timeframe allows, before sending the result PDU, the responder may send an ACK PDU. The initiator is unable to send the invoke message again needlessly [82].

6.10 A DATA-DRIVEN DEFENSE APPROACH TOWARD SECURE IoT DEVICES

To ensure that all feasible efforts are taken to address conceivable vulnerabilities, effective security measures for any IoT system involve a holistic view of the overall design and a data-driven strategy [72, 73]. When interacting with any outside impact,

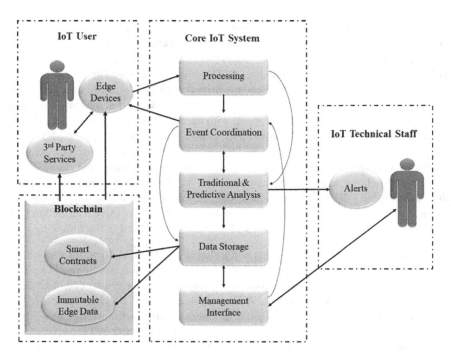

FIGURE 6.6   For IoT systems, a data-driven defense framework.

this entails taking into account all system components, their interrelationship as a whole, and the desired behavior. This also entails constructing the system in such a way that sufficient operational data is available for analyzing system integrity, performance, and mistakes [74]. The architecture suggested here defines a generalized system from which to construct a secure IoT system, and it is accompanied by essential principles aimed at promoting safe IoT system design and operation [83]. It is worth emphasizing that the data-driven IoT defensive framework, as shown in Figure 6.6, is being considered in the context of modern clustered computer big data systems. Modern networks operate at high speeds, necessitating parallel and concurrent computing. Because this is an important design element, it is mentioned openly rather than implied [83].

a. ***Processing:*** The processing subsystem receives input from edge devices and manages edge device validation and authorization, as well as governs their capacity to communicate within the system. The event coordination subsystem receives authorized traffic. The processing output is also sent to the analytics engine for further analysis. The key step of validating edge devices and end-user inputs is critical to the overall success of the system; thus, it must be carefully executed to ensure the data entering the system is trustworthy [84].

b. ***Event Coordination:*** Authorized traffic from edge devices is forwarded to this location for processing. Operations with a high risk of failure are redirected to edge devices. The analytics engine receives transactional and operational events, which are stored in data storage. This subsystem serves as the system's logical hub, facilitating communication between all other subsystems and ensuring that all transactions are completed or flagged as erroneous [56].

c. ***Analytics Engine:*** The analytics engine analyses the events received from event coordination. Some of these events are evaluated by prediction models that have been pre-trained, which will be addressed in the next subsection. Events and predictive models may produce warnings for technical staff to analyze and maybe act on in some circumstances. To handle modifications or idea drifts, models are continually retrained on new data [84]. This subsystem aggregates events in the logs and KPIs to offer the necessary data to system workers, as well as to encourage a better knowledge of system operation so that management tasks can be well informed, to take suitable steps in the system's day-to-day operation [56].

d. ***Data Storage:*** A storage subsystem stores all of the system's necessary data and provides it to the core components, edge devices, and third-party services as needed to assure functionality. To ensure data security, data within the system is encrypted at rest. Edge and third-party services interface with the data storage subsystem using one of two methods: A sequence of transactions with the processing and event coordination subsystems, or a publicly accessible blockchain subsystem.

e. ***Management Interface:*** The Management subsystem provides a user interface for technical employees to engage with the system for purposes such as configuration modifications, threat mitigation, and other operational activities required to ensure system fidelity and continuing operation. This subsystem is expected to be intimately integrated with the analytics engine, allowing system workers to take action based on the analytics engine's system dashboards, reports, and alarms [85].

f. ***Blockchain:*** The blockchain subsystem provides a publicly accessible data set for edge devices and third parties, enforcing the architecture and configuration of the IoT system and governing interactions with third-party systems and users. The blockchain's immutability makes it a suitable repository for IoT system data that has to be publicly accessible to edge devices. This immutable nature also enables the establishment of smart contracts that can safely and equitably oversee and manage the behaviors of edge devices and third-party services. Smart contracts can also be used to respond to events that are triggered by third-party vendor service level agreements [86]. Furthermore, licensing management for edge devices is an excellent candidate for smart contract management. The interface between the consumer and the core IoT system is provided via edge devices. Edge devices also give the IoT consumer a kind of utility or service. Edge devices are subject to manipulation and reverse engineering because of their physical availability, thus, extra care is necessary to build secure communication protocols and immutable behavior supplied by smart contracts embedded in the blockchain ledger. Immutable data encoded in the blockchain ledger is also used by edge devices [85].

g. **Third-party Services:** Third-party services are those that are not under the control of the IoT provider. These services interact with edge devices, and their behavior and configuration are stored as immutable data in the blockchain ledger or the context of a smart contract and may be accessed as needed. A commercial domain name provider that offers the system information required by the IoT system's edge devices, or an internet service provider that provides IP traffic transit, are examples of third-party services [86].

6.11   REQUIREMENTS FOR DATA-DRIVEN SECURITY AND PRIVACY IN WBANS

For wireless body area networks (WBAN's) system security, two crucial components are the security and privacy of patient-related data-driven information. Data-driven security refers to the fact that data-driven information is securely kept and transported, while data privacy refers to the fact that data-driven content may only be accessed by those who have been granted permission to view and use it. The security criteria are outlined next [87]. A distributed healthcare application scenario exemplifies the security requirements of WBANs. Assume Peter has an injury while traveling far from home. The emergency paramedic collects Peter's profile and medical records by reading his implanted RFID tag, and a WBAN of wearable medical sensors is built and linked to Peter. Later, numerous healthcare personnel can acquire real-time vital sign readings from the WBAN to provide better medical care. A nurse, for example, uses his WBAN to inquire about Peter's health status before uploading an electronic report to the local server in Peter's room. On Peter's PDA, an initial access policy (AP) has been established to control who has access to his WBAN medical data-driven. [87]. As illustrated in Figure 6.7, there are two tiers:

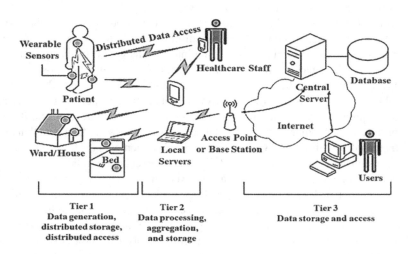

FIGURE 6.7   The general architecture of the WBAN [87].

Tier 1 and Tier 2. The information is either maintained in the WBAN for distributed, local access or transferred from the WBAN to Tier 3 medical databases for centralized, remote access. Patients, doctors, nurses, support workers, scientists, and insurance companies are all possible users of the WBAN's patient-related data.

The AP automatically adjusts many settings, including accommodating receptionists, doctors, and nurses. Peter can make changes to the AP at any time; for example, his sensitive AIDS data is only shared with his nurses, not doctors. Medical data is frequently stored and accessed in a distributed fashion. Different types of data-driven monitoring can be saved in different sensor nodes; however, the data can only be saved locally in Peter's WBAN until he gets to a site with wireless Internet connectivity. Direct local access to Peter's WBAN and local servers' cached data-driven information enables in-the-moment diagnosis by allowing freshly created data to be seen immediately. A logical question arises here: How can the security of distributed patient-related data be ensured from storage to access? Before going into the security of dispersed data-driven access, there are a few things to keep in mind, let's look at the threats that distributed data-driven information in the WBAN faces [87]. The WBAN frequently functions in contexts where multiple persons (e.g., medical personnel) have open access, which allows attackers to enter. Because of the open wireless channel, data can be intercepted, changed, and injected. These dangers have previously been thoroughly investigated in the literature. Visualize the threats from the device's perspective because this article focuses mostly on data-driven information and access.

   a. ***Device Compromise Threats:*** Because sensor nodes in a WBAN are often easy to capture, they are prone to compromise. If a piece of data-driven information is directly encrypted and stored in a node with its encryption key, data-driven leakage will occur if the node is breached. Local servers may also be untrustworthy because bad individuals are attempting to hack into them to obtain patients' personal information. They can either launch an attack over the Internet or just travel to a patient's room and wait for an opportunity to physically hack a local server.
   b. ***Threats Posed by Network Dynamics:*** The WBAN is a very dynamic system. Nodes may join and exit the network regularly due to unintentional failure or malicious activity. Due to a lack of power, nodes may die off. Attackers may readily place false sensors to impersonate genuine ones, and they could remove valid nodes on purpose. Because of network dynamics, patient-related data that is not well stored in more than one node may be quickly lost. Due to a lack of authentication, fake data could be injected or viewed as legitimate [87].

6.12   DISTRIBUTED DATA-DRIVEN SECURITY

Data-driven security is a type of decision-making that is based on data. Data-driven decision-making is a method of gathering information based on measurable objectives and then extracting facts, patterns, correlations, insights, and knowledge from the information. This information is then applied to the development or revision of processes, activities, systems, policies, and strategies that benefit the data/system owner. A company, a private or public organization, or a government body can be the data's owner. The objectives can be to boost corporate profitability, improve consumer happiness with an organization's products or services, increase student retention in schools and universities, or give a foundation for system design, administration, and enhancements in general.

Data-driven security strives to improve the security measures of an application, a system, or a whole business or environment by utilizing acquired data (general and specialized security data). The effective use of security data can aid in reducing the likelihood of security incidents, lowering security risks, detecting and mitigating security incidents, and improving the availability and quality of services. Data-driven security is a set of concepts that work together to make security decisions based on data that have been collected and analyzed rather than guesswork [57, 88, 89]. The data sets collected from that system and its surrounding and operational environment form the foundation of data-driven security for that system. Data clearing, filtering, anonymization, aggregation, organization, storage, exploration, analysis, discovery, and knowledge creation are some of the processes that these data sets may go through. This will result in the generation of security actions to increase the system's level of security [90].

Confidentiality is one of the required distributed data-driven security measures to avoid data-driven leaking during storage periods. Patient-related data must always be kept confidential at a node or local server. Data-driven confidentiality should be resistant to device compromise attacks; that is, if an attacker compromises one node, he or she should be unable to access the data on that node or elsewhere. [87]. Another important issue is the assurance of dynamic integrity, for which patient-related data is critical in WBANs, and changing it might have severe repercussions. As a result, data integrity will be safeguarded in real-time. Not only will able to detect data-driven changes at end users but will be also verified and detect them during storage periods, allowing us to detect potentially problematic data-driven modifications in advance and alert the user [57]. Moreover, dependability is also required as a major worry in WBANs, because failure to obtain proper data might be life-threatening. Fault tolerance, or the ability to retrieve patient-related data even in the event of Byzantine node failure or malicious alterations, is necessary to combat the dangers posed by network dynamics.

6.12.1 DISTRIBUTED DATA-DRIVEN ACCESS SECURITY IN WBANs

In controlling data-driven access at a finer level, access control for patient-related data must be established in WBANs to ensure that unauthorized parties do not gain confidential information. Doctors, support staff, pharmacies, and other organizations could study Peter's medical data in the application scenario their services were outlined to improve. For example, if an insurance company learns of Peter's illness, it may treat him unfairly by charging him a high health insurance premium. As a result, to declare and enforce different access privileges for distinct users, it is necessary to define a fine-grained access policy. The fine granularity of the data-driven access policy, which distinguishes between each portion of the patient's data and each user role, is referred to as fine-grained. For example, "personal identifying information such as patient profile shall not be divulged to insurance companies," or "doctors are only entitled to study the medical data of patients they are treating, not that of other patients [89]."

Scalability is required because there are so many users of patient data; the distributed access control technique should be scalable in the following ways. First, having a low administration overhead for access controls that can be simply set up and adjusted. Second, having a low overhead in terms of processing and storage will be demonstrated in further detail in the following section. Flexibility is one of the most important needs is that the patient can designate APs for his data based on his personal preferences. More crucially, APs will be enabled to dynamically respond to situations affecting patients, such as time, location, or unique occurrences. When a medical emergency arises, for example, an available doctor who is not on the allowed list can be given on-demand authorization to access Peter's monitoring data. The inability or unwillingness to adjust access rules could put a patient's safety in jeopardy. Additional security requirements include accountability, revocability, and nonrepudiation, which are outlined in Table 6.3. In general, to prevent fake data injection and DoS attacks, authentication is a security service that verifies a user's identity before data access. It is also necessary for secure data transmission within the WBAN. Because authentication is not the major focus of this work, it is discussed only when it is necessary [87].

TABLE 6.3
Major Security Requirements for Critical Data-Driven Security & Privacy

Security Requirements	Description	
Data Storage Security	Confidentiality	Critical related WBAN data must be kept confidentiality during the storage process Robustness against user collision
	Dynamical Integrity Assurance	Data must be modified illegally during the storage period, checked, and detected dynamically
	Dependability	Data must be retrievable in case of failure
Data Access Security	Access Control (Privacy)	Preventing unauthorized access to the critical data in the network
	Accountability	Identify and hold accountable users who are privileged to carry out unauthorized actions
	Revocability	The privilege of users should be revoked immediately if they are identified as behaving maliciously
Others	Authentication	Data sender must be authenticated, and any injection of outside data to the network must be prevented
	Availability	Data must be accessible even under the DoS attack

6.12.2 PRACTICAL CHALLENGES THAT ARE DIFFICULT TO SOLVE

Many major problems must be overcome to achieve the following objectives in WBAN, most of which stem from efficiency and practicality concerns. These constraints restrict the solution area and must be carefully considered while constructing data-driven security and privacy safeguards in WBANs.

   a. *Security vs. Efficiency:* There is an on-going battle. Because of resource constraints as well as application requirements, high efficiency is necessary for data-driven security in WBANs. Wearable sensors are often small and have a limited power supply, limiting their computation and storage capabilities. As a result, the cryptographic primitives used by the sensor nodes should be as light as possible in terms of computation time and storage overhead. Otherwise, the nodes' power and storage space may be drained very quickly. In addition, if the authentication procedure is not fast enough, a DoS attack might potentially overload the entire WBAN [87].
   b. *Security vs. Safety:* Whether or not data-driven can be accessed anytime it is needed could be a safety issue for patients [91]. Data access controls that are too rigid and inflexible may hinder legitimate medical personnel from accessing medical information promptly, especially in emergencies where the patient is unconscious and unable to reply. A sloppy access control scheme, on the other hand, gives malicious attackers back doors. It is difficult to maintain data-driven security and privacy while allowing for a wide range of access. When there is network coverage, some methods allow greater authentication of the user by connecting to the authority; Weaker or no authentication is used when the infrastructure is not available, such as during disaster response [92]. Their approach can be considered the first step towards solving the security and safety dilemma.
   c. *Security and Usability are at Odds:* Because patients who are not experts could be their operators, the gadgets must be simple to use and fail-safe. The data-driven security systems' setup and control processes will entail few intuitive human interactions because they are patient-related. Device-pairing approaches, for example, can be used to establish a secure connection between all nodes in a WBAN for data-driven exchange.
   d. *Interoperability Requirement:* Patients can purchase sensor nodes from a variety of manufacturers, preventing the advanced sharing of cryptographic materials. It is challenging to set up data-driven security procedures that use a variety of devices and require the fewest settings and efforts.

6.12.3 DISTRIBUTED DATA-DRIVEN THAT IS SECURE AND DEPENDABLE

Data confidentiality, dependability, and integrity are the three most important requirements for distributed data-driven security in WBANs. To provide redundancy and increase data dependability, error-correcting code techniques might be used. A study by S. Chessa et al., proposed a secure distributed data-driven and sharing mechanism based on redundant residue number system (RRNS) [93]. An integer in

RRNS is represented as $h + r$ moduli, with the extra r moduli being omitted. In the proposed architecture, a source node S distributes a file F to n other nodes. S chooses $n = h + r$ moduli at random, computes F's residue vector, and allocates each file share to its storage node. An authorized node must collect enough residue from storage nodes to retrieve the original file. The presented method improves dependability since the RRNS can absorb up to sr data share erasures and up to reach corruption and data can be reconstructed using any h of the remaining accurate shares. For the same reason, resistance to compromise of up to h nodes is achieved. To ensure privacy, the data-driven sharing is encrypted using the public keys of approved storage nodes. However, data integrity is not guaranteed when the number of errors exceeds the detecting capability. Distributing public keys to sensor nodes for interoperability is likewise a bad idea. In terms of efficiency, the length expansion ratio for each file is $(h + r)/h$. Keeping a potentially large collection of moduli in the buffer of a sensor node, on the other hand, would be inefficient for WBANs. A node in a WBAN should be able to dynamically check the integrity of data-driven sharing in other nodes before a user obtains data from another node. The use of message authentication codes alone results in a large increase in storage requirements.

Q. Wang et al., proposed a reliable and secure distributed data-driven strategy. The encrypted data is divided into n data-driven shares, each containing a data-driven block generated by (n, k)-erasure coding and a share of the secret key obtained through (n, k)-secret sharing [94]. The data-driven shares are then distributed to n neighbor nodes for storage. Each other storage node computes and broadcasts an algebraic signature on one data-driven share to perform a dynamic integrity check, allowing the checking node to validate the integrity by comparing the other nodes' signatures to its own. Any changes to the data will be detected in real-time in this manner. In the proposed approach, data-driven confidentiality, reliability, and dynamic integrity are all guaranteed at the same time. For WBANs, this could be inconvenient because we would prefer the local server to double-check the data's integrity once it has been collected. Because a more powerful attacker may attempt to destroy/erase valuable medical data, i.e. vital sign readings in WBANs, a higher level of dependability, known as data survivability, may be necessary. A study presented by R. Di Pietro et al., solved the data-driven survivability problem in wireless sensor networks. Each round can only compromise a subset of sensor nodes, and the attacker is assumed aware of the source of the target data [95]. The basic premise behind the defense is to constantly move data from one sensor node to the next, making it more difficult for the opponent to "catch" it.

6.13   MOVING TOWARD DATA-DRIVEN SECURITY

Yet, if a lot of ground is followed, a lot may be noticed, and it is hoped that there will be a realization that the data contains wisdom. When moving security practices to a data-driven perspective, the use of a "gold prospecting" approach rather than an "oil drilling" approach is advocated. This means you should not become stuck on a single focus (or data source) at a time. Instead, roll up your slacks, enter the data stream, and simply explore and learn as much as possible. After you have grasped what is data-driven, you can begin asking and answering interesting questions that will begin to make a difference. That distinction is the focus of the final section. On a personal level, the first part is about encouraging yourself or others you work with to adopt a data-driven mindset. The second half of the section will cover how to make the move to a data-driven security program at your firm. [96].

6.13.1 MOVING HUMANS TOWARD DATA-DRIVEN SECURITY

The data-driven venn diagram shown in Figure 6.8 is a significantly modified and simple graphic that helps rapidly assess where you are on the road to data-driven security. Each important component is examined in this chapter, as well as some of their relationships [97]. The purpose is to help to identify areas that are currently deficient. It is not a requirement that strength is present in all the previously mentioned major areas, but any weaknesses should be addressed. The following sections illustrate some of the weaknesses related to data-driven security.

a. **The Hacker**

Because it has been usurped by the news media and used by advertising organizations, the term "hacker" has a lot of ambiguity attached to it [98]. "Hacker" is redefined in the context of a security data scientist to include abilities such as: A) The ability to manipulate computers using code, whether by scripting in Python or full-fledged C programming; B) An understanding of a variety of data formats, as well as how to slice, dice, and bend them to your will; C) Being able to think critically, logically, scientifically (i.e., without leaping to conclusions), and algorithmically by dividing a problem into its constituent components; and D) Using visuals, tables, charts, or simply an excellent, old-fashioned word collection to communicate your work.

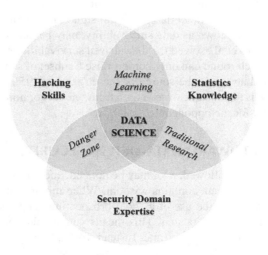

FIGURE 6.8 The data-driven Venn diagram.

b. **The Manager of Data**

When it comes to data formats, security specialists must deal with anything from net flow captures to full packet capture (PCAP) dumps, as well as virtually every log type imaginable. The "MongoDB" sample of the IronPort log file is an illustration of your data universe's "imperfection." Although that log file has a lot of helpful information, it's in a format that you'll have to read and convert before you can use it [99]. The only way to become proficient at it is to practice repeatedly, accumulating reusable code and techniques along the road to save time in the future.

6.13.2 MOVING ORGANIZATION TOWARD DATA-DRIVEN SECURITY

It has probably now been realized that becoming data-driven entails more than merely starting up R or Python and throwing data at it. Developing a data-driven mindset is a gradual process that will change the way you and others in your business see the world. The value will not be apparent right away. Rather, the worth will grow over time, punctuated by bursts of brilliance. The following are some of the components of a good data-driven program within any organization [58]. Ask questions that can be answered objectively. Locate and collect pertinent information. In addition, iteration is a great way to learn. Because the first two elements generate a chicken-and-egg situation, the most challenging portion of the change is getting started. In some cases, questions will be asked for which you already have data, but your goal was just to gather the information that will assist you in answering those questions. Do not be concerned; you can improve both through repetition.

a. **Locate and Gather Pertinent Information**

Data gathering and asking effective questions are inextricably linked. You'll need data to answer questions, but may have no desire to collect information that is not be needed. Which comes first, orange color or fruit? Having an idea of the data available only through logs from the proxy and firewall, as well as logs from server authentication and even data from a business ticket system, are all useful places to start. Always start by formulating some practical issues that the data can address [100]. There may be a need to review the questions during data collection to answer the queries, and then learn more about the data and modify it again. Be prepared to collaborate with others to collect data. There is a high chance that not all the data that will be needed will be accessible. This is why it is important to have executive sponsorship. In the case of corporate responsibility, data sharing must be made a priority [101].

b. **Iterative Learning**

A standard waterfall project plan cannot be used to create a data-driven security program, where actions are planned and then executed one by one. As shown in Figure 6.9, it is a more iterative process, and the transition from inquiry to conclusion can quickly become a maze. Each data source comes with its own set of challenges and opportunities. Setbacks and obstacles become as much part of the endeavor as they become a success, and

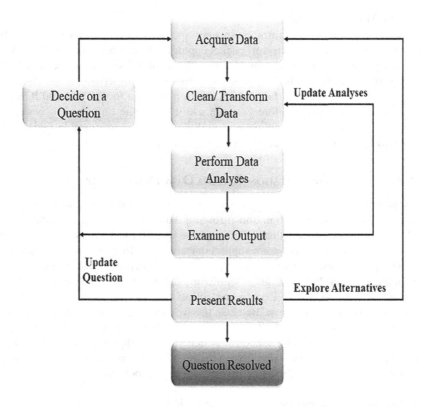

FIGURE 6.9 The data-driven workflow.

repetition becomes the mantra. However, do not get discouraged. Relapses will become less frequent, and each one will be a learning experience [102].

One of the most important lessons that will be learned early on is the necessity of data quality and the benefits of including repeatability. It will not be long until the extractor checks out and it has revealed that the date variable has been corrupted, a field has been clipped, or another normal occurrence necessitates repeating the entire procedure. Therefore, in addition to automating the extraction, transformation, and loading tools, data validations must be performed frequently. It's important to be aware that data integrity has been compromised long before the final report is generated [103].

6.13.3 BUILDING A REAL-LIFE SECURITY DATA-DRIVEN SYSTEM

In the past, it was difficult for the user at first to resist the great security concerns that could face data systems, until the Hadoop group provides a way to import every log from every secured system into a massive data-driven system [104]. In general, data is treated as a way to make strategic decisions about network security in addition to cybersecurity. Therefore, there is a need to achieve the security aspect in the use of resources according to a system that meets the requirements of security and privacy. For data-driven systems, an unprecedented view of existing security standards must be developed based on the security performance data of hundreds of thousands of global organizations. It is also very important to identify gaps in security performance according to standards and a data-driven processing plan. Some steps are followed to build a data-driven security system, so software such as JavaScript, SQL (MariaDB), and NoSQL (MongoDB & Redis) can help to find new methods for data acquisition, cleaning, and storage known as the process of organizing data and formalizing queries and schemas to the structure [105].

For big data, the process of collecting data, especially those related to IoT networks, needs a mechanism that works to protect the confidentiality of devices and share the huge volume of their data. Therefore, it is necessary to consider the infrastructure of IoT and determine the capabilities of storage systems, especially cloud-based, that organizations can use easily [106]. Broadly speaking, data-driven security is built into an environment by defining all decisions and operations through data, and enhancing security software behavior by balancing the following factors: Selecting and deploying effective security measures, operating within budget, and minimizing liability exposure.

6.14   INTELLIGENT SECURITY DATA-DRIVEN FUTURE ASPECTS

Cybersecurity is a set of technologies and procedures aimed at protecting computers, networks, programs, and data from assaults, damage, and unauthorized access. The change is being driven by data science, a fundamental feature of artificial intelligence (AI), and ML, a core part of AI that may play a key role in discovering insights from data. Data science is at the vanguard of a new scientific paradigm, and ML has the potential to significantly alter the cybersecurity landscape. Cybersecurity driven by data science (CDS) has the ultimate goal of empowering security professionals to make data-driven, well-informed judgments [105]. CDS is a partial paradigm change from old well-known security solutions such as firewalls, user authentication, access control, encryption systems, and so on, which may or may not be effective in today's cyber sector needs. The issues are that a few competent security analysts usually address them statically. However, since the frequency of cybersecurity incidents in the various formats listed previously continues to rise over time, traditional solutions have proven ineffective in managing such cyber dangers [105, 106].

As a result, a large number of complex assaults are developed and disseminated swiftly across the Internet. Although several researchers use various data, analysis, and learning techniques to build cybersecurity models, as discussed in ML tasks in the cybersecurity section, a comprehensive security model based on the effective discovery of security insights and the latest security patterns may be more useful. To address this issue, there is a need to create more resilient and efficient security systems that can respond to attacks and change security rules in time to mitigate them intelligently [107]. To reach this purpose, it is necessary to analyze a large amount of important cybersecurity data collected from many sources, such as network and system sources, and to identify insights or correct security policies in an automated

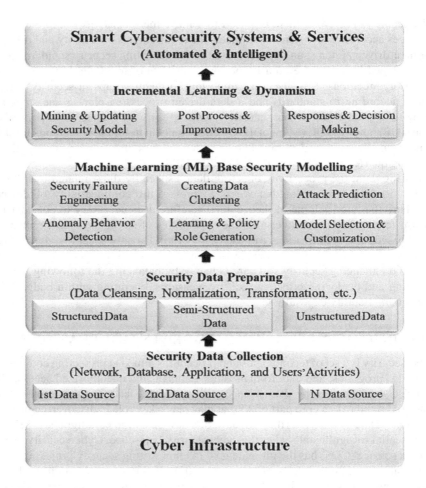

FIGURE 6.10 A generic multi-layered framework for smart cybersecurity based on machine learning techniques.

manner with minimal human interaction. ML is a subset of AI that is strongly linked to computational statistics, data mining and analytics, and data science, with a focus on teaching computers to learn from data [106].

ML models, for example, are often made up of a set of rules, procedures, or sophisticated "transfer functions" that may be used to uncover interesting data patterns or detect or anticipate behaviors [84], which could be useful in the field of cybersecurity. Some of the strategies that can be used to solve ML problems, as well as how they relate to cybersecurity issues, are shown in Figure 6.10. The most crucial task for an intelligent cybersecurity system is to provide an efficient framework that allows for data-driven decision-making. Advanced data analysis based on ML techniques should be included in such a framework to be able to reduce these concerns and provide automated and intelligent security services.

As a result, a well-designed security framework for cybersecurity data and experimental evaluation is a critical direction as well as a significant issue. ML-based security modeling is the fundamental step in the CDS process when insights and knowledge are collected from data. The emphasis on ML-based modeling is because of that, ML techniques have the potential dramatically change the cybersecurity landscape. The discovery and analysis of security traits or attributes, as well as their patterns in data, is of great importance to derive security insights.

Greater knowledge of data and machine learning-based analytical models leveraging a vast amount of cybersecurity data can be beneficial in achieving the goal. Depending on the answer, several ML activities might be included in this model building layer [105, 108]. These include security feature engineering, which is in charge of transforming raw security data into relevant features that effectively reflect the underlying security problem to data-driven models. As a result, depending on the security data characteristics, this module may include feature transformation and normalization, feature selection by considering a subset of available security features based on their correlations or importance in modeling, or feature generation and extraction by creating new brand principal components. The chi-squared test, analysis of variance test, correlation coefficient analysis, feature importance, discriminant, and principal component analysis, or singular value decomposition, for example, can all be used to analyze the significance of security features to perform security feature engineering tasks [57].

6.15 CONCLUSION

It is clear that there is a direct relationship between the volume of information and the increase in security risks. For this reason, security in wireless networks may be addressed, and all parts responsible for security within the network, through the seven layers from the application layer to the physical layer are known, and each function of each layer in network security is known. Then, the types of vulnerabilities and potential attacks on the network were known, whether through layers or in general. The definitions of attacks and threats in general on wireless network technology were discussed. Some suggested solutions and applications used to protect wireless networks, such as WPA and WPA2. After that, the requirements for data-driven security and privacy in WBANs were discussed as well as the requirements for distributed data-driven security.

Many topics related to person and organization protection were discussed as well as dealing with data streams through which protection was achieved. In addition, the discussion deals with data in the wireless network, whether searching for data, collecting data, asking questions, or answering them. Be careful to not fall into any gap that may lead to threatening the wireless network, data, or both. Much knowledge was gained from the pages of this chapter and recognize that there is no requirement to do everything at once. The transition to a data-driven lifestyle can be achieved by combining hacking skills, an understanding of the field, and statistics. The path to data-driven security software for a company will be shown when those skills are combined with the art of asking the right questions and gathering data to answer those questions. To get results, there is no need to immediately put everything into place. Over time, the iterative strategy should offer benefits and help adapt to inevitable problems. Start slow, do everything, then try again and report your results.

REFERENCES

1. M. Bostock, V. Ogievetsky and J. Heer, "D³ Data-Driven Documents," in *IEEE Transactions on Visualization and Computer Graphics*, vol. 17, no. 12, pp. 2301–2309, Dec. 2011, doi: 10.1109/TVCG.2011.185.
2. A. A. Muthana and M. M. Saeed, "Analysis of user identity privacy in LTE and proposed solution," International Journal of Computer Network and Information Security, vol. 9, no. 1, 2017.
3. Saeed, M.M., Hasan, M.K., Obaid, A.J., Saeed, R.A., Mokhtar, R.A., Ali, E.S., Akhtaruzzaman, M., Amanluo, S., Hossain, A.K.M.Z.: A comprehensive review on the users' identity privacy for 5G networks. *IET Communications* Vol. 16, 2022.
4. L. Deri *et al.*, "Beyond the Web: Mobile WAP-based management," *Journal of Network and Systems Management*, vol. 9, no. 1, pp. 15–29, 2001.
5. M. M. Saeed, R. A. Saeed, and E. Saeid, "Preserving privacy of paging procedure In 5th G using identity-division multiplexing," in *2019 First International Conference of Intelligent Computing and Engineering (ICOICE)*, 2019, pp. 1–6: IEEE.
6. T. Sobh *et al.*, "Wired and wireless intrusion detection system: Classifications, good characteristics and state-of-the-art," Computer Standards & Interfaces, vol. 28, no. 6, pp. 670–694, 2006.
7. M. K. Afzal, M. H. Rehmani, Y. B. Zikria, and Q. Ni, "Data-driven intelligence in wireless networks: Issues, challenges, and solution," *Transactions on Emerging Telecommunications Technologies*, vol. 30, p. e3722, 2019.
8. R. Ali, I. Ashraf, A. K. Bashir, and Y. Zikria, "Reinforcement-Learning-Enabled massive Internet of Things for 6G wireless communications," in *IEEE Communications Standards Magazine*, vol. 5, no. 2, pp. 126–131, 2021.
9. R. Ali, Y. B. Zikria, S. Garg, A. K. Bashir, M. S. Obaidat, and H. Kim, "A federated reinforcement learning framework for incumbent technologies in beyond 5G networks," in *IEEE Network*, vol. 35, no. 4, pp. 152–159, 2021.
10. A. Irshad *et al.*, "A secure demand response management authentication scheme for smart grid," Sustainable Energy Technologies and Assessments, Elsevier, vol. 48, p. 101571, 2021.
11. S. A. Chaudhry *et al.*, "An anonymous device to device access control based on secure certificate for internet of medical things systems," Sustainable Cities and Society, Elsevier, vol. 75, p. 103322, 2021.
12. S. A. Chaudhry, A. Irshad, K. Yahya, N. Kumar, M. Alazab, and Y. Zikria, "Rotating behind privacy: An improved lightweight authentication scheme for cloud-based IoT environment," ACM Transactions on Internet Technology, vol. 21, no. 3, pp. 1–19, 2021.
13. S. A. Chaudhry, K. Yahya, M. Karuppiah, R. Kharel, A. K. Bashir, and Y. Zikria, "GCACS-IoD: A certificate based generic access control scheme for Internet of drones," Computer Networks, Elsevier, vol. 191, p. 107999, 2021.
14. S. Islam, J. Lloret, and Y. Zikria, "Internet of Things (IoT)-based wireless health: Enabling technologies and applications," Electronics, MDPI, vol. 10, p. 148, 2021.
15. Z. Ali, S. A. Chaudhry, K. Mahmood, S. Garg, Z. Lv, and Y. Zikria, "A clogging resistant secure authentication scheme for fog computing services," Computer Networks, Elsevier, vol. 185, p. 107731, 2021.
16. M. Rana *et al.*, "A secure and lightweight authentication scheme for next generation IoT infrastructure," Computer Communications, Elsevier, vol. 165, pp. 85–96, 2021.

17. R. Majeed, N. A. Abdullah, I. Ashraf, Y. B. Zikria, M. F. Mushtaq, and M. Umer, "An intelligent, secure, and smart home automation system," *Scientific Programming, hindawi*, vol. 2020, 2020.

18. M. Sohail *et al.*, "Trustwalker: An efficient trust assessment in Vehicular Internet of Things (VIoT) with security consideration, "sensors, MDPI, vol. 20, no. 14, p. 3945, 2020.

19. I. F. Akyildiz, W. Su, Y. Sankarasubramaniam, and E. Cayirci, "A survey on sensor networks," IEEE Communications Magazine vol. 40, no. 8, pp. 102–114, 2002.

20. E. Shi and A. Perrig, "Designing secure sensor networks," IEEE Wireless Communications, vol. 11, no. 6, pp. 38–43, 2004.

21. Z. Al-Salloum and S. Wolthusen, "Semi-autonomous link layer vulnerability discovery and mitigation dissemination," in *2009 Fifth International Conference on IT Security Incident Management and IT Forensics*, IEEE, pp. 41–53,2009.

22. A. Koubâa, R. Severino, M. Alves, and E. Tovar, "Improving quality-of-service in wireless sensor networks by mitigating 'hidden-node collisions," IEEE Transactions on Industrial Informatics, vol. 5, no. 3, pp. 299–313, 2009.

23. N. Jiang, Y. Deng, A. Nallanathan, X. Kang, and T. Quek, "Analyzing random access collisions in massive IoT networks," IEEE Transactions on Wireless Communications, vol. 17, no. 10, pp. 6853–6870, 2018.

24. Sahabul Alam ` and Debashis De, "Analysis of security threats in wireless sensor network," International Journal of Wireless & Mobile Networks (IJWMN) Vol. 6, No. 2, April 2014

25. D. Reed, "Applying the OSI seven layer network model to information security," white paper, SANS institute, vol. 1, p. 8, 2003.

26. C. Karlof and D. Wagner, "Secure routing in wireless sensor networks: Attacks and countermeasures," Proceedings of the First IEEE International Workshop on Sensor Network Protocols and Applications, vol. 1, no. 2-3, pp. 293–315, 2003.

27. A. Perrig, R. Szewczyk, J. D. Tygar, V. Wen, and D. Culler, "SPINS: Security protocols for sensor networks," Wireless Networks, springer, vol. 8, no. 5, pp. 521–534, 2002.

28. J. Newsome, E. Shi, D. Song, and A. Perrig, "The sybil attack in sensor networks: Analysis & defenses," in *Third international symposium on information processing in sensor networks, 2004. IPSN 2004*, 2004, pp. 259–268: IEEE.

29. Y. Hu, A. Perrig, and D. B. Johnson, "Packet leashes: A defense against wormhole attacks in wireless networks," in *IEEE INFOCOM 2003. Twenty-second Annual Joint Conference of the IEEE Computer and Communications Societies (IEEE Cat. No. 03CH37428)*, 2003, vol. 3, pp. 1976–1986: IEEE.

30. K. Al-Gharbi and S. Naqvi, "The use of Intranet by Omani organizations in knowledge management," International Journal of Education and Development using Information and Communication Technology (IJEDICT), vol. 4, no. 1, pp. 27–40, 2008.

31. M. Harsha, B. Bhavani, and K. Kundhavai, "Analysis of vulnerabilities in MQTT security using Shodan API and implementation of its countermeasures via authentication and ACLs," in *2018 International Conference on Advances in Computing, Communications and Informatics (ICACCI)*, 2018, pp. 2244–2250: IEEE.

32. P. K. Gupta, S. Mittal, P. Consul, and J. K. Jindal, "A review of different vulnerabilities of security in a layered network," Advances and Applications in Mathematical Sciences, volume 20, issue 2, pages 227–236, 2020.

33. A. A. Khwaja, M. Murtaza, and H. Ahmed, "A security feature framework for programming languages to minimize application layer vulnerabilities," *Security and Privacy*, vol. 3, no. 1, p. e95, 2020.

34. J. Eriksson, S. V. Krishnamurthy, and M. Faloutsos, "Truelink: A practical countermeasure to the wormhole attack in wireless networks," in *Proceedings of the 2006 IEEE International Conference on Network Protocols*, 2006, pp. 75–84: IEEE.

35. J. Mirkovic and P. Reiher, "A taxonomy of DDoS attack and DDoS defense mechanisms," *ACM SIGCOMM Computer Communication Review*, vol. 34, no. 2, pp. 39–53, 2004.

36. W. G. Halfond, J. Viegas, and A. Orso, "A classification of SQL-injection attacks and countermeasures," in *Proceedings of the IEEE international symposium on secure software engineering*, 2006, vol. 1, pp. 13–15: IEEE.

37. A. Hadid, N. Evans, S. Marcel, and J. Fierrez, "Biometrics systems under spoofing attack: An evaluation methodology and lessons learned," in *IEEE Signal Processing Magazine*, vol. 32, no. 5, pp. 20–30, 2015.

38. D. Guinier, "From eavesdropping to security on the cellular telephone system GSM," *ACM SIGSAC Review*, vol. 15, no. 2, pp. 13–18, 1997.

39. J. Slupska, "Safe at home: Towards a feminist critique of cybersecurity," Slupska, Julia, Safe at Home: Towards a Feminist Critique of Cybersecurity (May 1, 2019). St. Anthony's International Review 2019 no. 15: Whose Security is Cybersecurity? Authority, Responsibility and Power in Cyberspace, vol. 15, no. 1, pp. 83–100, 2019.

40. K. Rieck, T. Holz, C. Willems, P. Düssel, and P. Laskov, "Learning and classification of malware behavior," in *International Conference on Detection of Intrusions and Malware, and Vulnerability Assessment*, 2008, pp. 108–125: Springer.

41. P. Baecher, M. Koetter, T. Holz, M. Dornseif, and F. Freiling, "The nepenthes platform: An efficient approach to collect malware," in *International Workshop on Recent Advances in Intrusion Detection*, 2006, pp. 165–184: Springer.

42. D. Stiawan, S. Sandra, E. Alzahrani, and R. Budiarto, "Comparative analysis of K-Means method and naïve Bayes method for brute force attack visualization," in *2017 2nd International Conference on Anti-Cyber Crimes (ICACC)*, 2017, pp. 177–182: IEEE.

43. J. McCulloch and L. Schetzer, "Brute force: The need for affirmative action in the Victoria Police Forces," Australian Feminist Law Journal, vol. 1, no. 1, pp. 45–62, 1993.

44. T. Varady, R. R. Martin, and J. Cox, "Reverse engineering of geometric models—an introduction," Computer-Aided Design, vol. 29, no. 4, pp. 255–268, 1997.

45. E. J. Chikofsky and J. Cross, "Reverse engineering and design recovery: A taxonomy," IEEE Software, vol. 7, no. 1, pp. 13–17, 1990.

46. V. Raja and K. J. Fernandes, *Reverse Engineering: An Industrial Perspective*. Springer Science & Business Media, 2007.

47. H. A. Müller, J. H. Jahnke, D. B. Smith, M.-A. Storey, S. R. Tilley, and K. Wong, "Reverse engineering: A roadmap," in *Proceedings of the Conference on the Future of Software Engineering*, 2000, pp. 47–60.

48. L. A. Hughes and G. DeLone, "Viruses, worms, and trojan horses: Serious crimes, nuisance, or both?," *Social Science Computer Review*, vol. 25, no. 1, pp. 78–98, 2007.

49. E. Karafili, L. Wang, and E. Lupu, "An argumentation-based reasoner to assist digital investigation and attribution of cyber-attacks," Forensic Science International: Digital Investigation, vol. 32, p. 300925, 2020.

50. Z. Bakhshi, A. Balador, and J. Mustafa, "Industrial IoT security threats and concerns by considering Cisco and Microsoft IoT reference models," in *2018 IEEE Wireless Communications and Networking Conference Workshops (WCNCW)*, 2018, pp. 173–178: IEEE.

51. M. A. Al-Garadi *et al.*, "A survey of machine and deep learning methods for Internet of Things (IoT) security," IEEE Communications Surveys & Tutorials, vol. 22, no. 3, pp. 1646–1685, 2020.

52. K. Bobrovnikova, S. Lysenko, P. Gaj, V. Martynyuk, and D. Denysiuk, "Technique for IoT Cyberattacks Detection Based on DNS Traffic Analysis," in *IntelITSIS*, 2020, pp. 208–218.

53. P. Aufner, "The IoT security gap: A look down into the valley between threat models and their implementation," International Journal of Information Security, vol. 19, no. 1, pp. 3–14, 2020.

54. S. M. Loo and L. Babinkostova, "Cyber-physical systems security introductory course for STEM students," 2020 ASEE Virtual Annual Conference Content Access, 2020.

55. M. Nawir, A. Amir, N. Yaakob, and O. B. Lynn, "Internet of Things (IoT): Taxonomy of security attacks," in *2016 3rd International Conference on Electronic Design (ICED)*, 2016, pp. 321–326: IEEE.

56. P. Wang, L. T. Yang, X. Nie, Z. Ren, J. Li, and L. Kuang, "Data-driven software defined network attack detection: State-of-the-art and perspectives," Information Sciences, Elsevier, vol. 513, pp. 65–83, 2020.

57. D. B. Rawat, R. Doku, and M. Garuba, "Cybersecurity in big data era: From securing big data to data-driven security," IEEE Transactions on Services Computing, vol. 14, Issue. 6, 2019.

58. M. Mohsin, Z. Anwar, F. Zaman, and E. Al-Shaer "IoT Checker: A data-driven framework for security analytics of Internet of Things configurations," Computers & Security, elsevier, vol. 70, pp. 199–223, 2017.

59. B. Yang, and M. Yang, "Data-driven network layer security detection model and simulation for the Internet of Things based on an artificial immune system," *Neural Computing and Applications*, vol. 33, no. 2, pp. 655–666, 2021.

60. M. Waliullah and D. Gan, "Wireless LAN security threats & vulnerabilities," International Journal of Advanced Computer Science and Applications(IJACSA), vol. 5, no. 1, 2014.

61. D. Welch and S. Lathrop, "Wireless security threat taxonomy," in *IEEE Systems, Man and Cybernetics Society Information Assurance Workshop*, 2003, pp. 76–83: IEEE.

62. Suroto, WLAN Security: Threats and Countermeasures, International Journal on Informatics Visualization, vol.2, no.4, 2018.

63. Y. Zahur and T. Yang, "Wireless LAN security and laboratory designs," *Journal of Computing Sciences in Colleges*, vol. 19, no. 3, pp. 44–60, 2004.

64. Z. Tao and A. Ruighaver, "Wireless intrusion detection: Not as easy as traditional network intrusion detection," in *TENCON 2005-2005 IEEE Region 10 Conference*, 2005, pp. 1–5: IEEE.

65. A. Sharadqeh, A. Alnaser, O. Heyasat, A. Abu-Ein and H. Hatamleh, "Review and measuring the efficiency of SQL injection method in preventing e-mail hacking," International Journal of Communications, Network and System Sciences, vol. 5, no. 6, pp. 337–342, 2012. doi: 10.4236/ijcns.2012.56044.

66. R. A. Saeed, M.M. Saeed, R.A. Mokhtar, H. Alhumyani, and S. Abdel-Khalek, "Pseudonym mutable based privacy for 5G User identity," *Computer Systems Science & Engineering*, vol. 39, no. 1, pp. 1–14, 2021.

67. S. Yi, P. Naldurg, and R. Kravets, "Security-aware ad hoc routing for wireless networks," in *Proceedings of the 2nd ACM international symposium on Mobile ad hoc networking & computing*, 2001, pp. 299–302.

68. L. Zhou and Z. Haas, "Securing ad hoc networks," IEEE Network, vol. 13, no. 6, pp. 24–30, 1999.

69. Y.C. Hu, A. Perrig, and D. Johnson, "Wormhole attacks in wireless networks," IEEE JOURNAL ON SELECTED AREAS IN COMMUNICATIONS, vol. 24, no. 2, pp. 370–380, 2006.

70. J. T. Isaac, S. Zeadally, and J. Camara, "Security attacks and solutions for vehicular ad hoc networks," International Journal on AdHoc Networking Systems, vol. 4, no. 7, pp. 894–903, 2010.

71. J. Sen, "Security and privacy issues in wireless mesh networks: A survey," in *Wireless networks and security*: Springer, 2013, pp. 189–272.

72. L. Hu and D. Evans, "Secure aggregation for wireless networks," in *Proceedings of Symposium on Applications and the Internet Workshops*, 2003, pp. 384–391: IEEE.

73. J. Zhu and J. Ma, "A new authentication scheme with anonymity for wireless environments," *IEEE Transactions on Consumer Electronics*, vol. 50, no. 1, pp. 231–235, 2004.

74. Y. M. Erten and E. Tomur, "A layered security architecture for corporate 802.11 wireless networks," in *Symposium on Wireless Telecommunications*, 2004, pp. 123–128: IEEE.

75. W. Alliance, "Wi-Fi protected access: Strong, standards-based, interoperable security for today's Wi-Fi networks," Wi-Fi Alliance, vol. 1, no. 2004, pp. 23–29, 2003.

76. P. Guo, J. Wang, X. H. Geng, C. S. Kim, and J. Kim, "A variable threshold-value authentication architecture for wireless mesh networks," journal of internet technology, vol. 15, no. 6, pp. 929–935, 2014.

77. N. B. Salem and J. Hubaux, "Securing wireless mesh networks," vol. 13, no. 2, pp. 50–55, 2006.

78. A. Kavianpour and M. C. Anderson, "An overview of wireless network security," in *IEEE 4th International Conference on Cyber Security and Cloud Computing (CSCloud)*, 2017, pp. 306–309: IEEE.

79. T. Heer, O. Garcia-Morchon, R. Hummen, S. L. Keoh, S. S. Kumar, and K. Wehrle, "Security challenges in the IP-based Internet of Things," vol. 61, no. 3, pp. 527–542, 2011.

80. P. Brutch and C. Ko, "Challenges in intrusion detection for wireless ad-hoc networks," in *2003 Symposium on Applications and the Internet Workshops, 2003. Proceedings.*, 2003, pp. 368–373: IEEE.

81. P. Goyal, V. Parmar, R. Rishi, and Management, "Manet: Vulnerabilities, challenges, attacks, application," vol. 11, no. 2011, pp. 32–37, 2011.

82. W. Stallings, *Cryptography and Network Security,* 4th Edition. Pearson Education India, 2006.

83. C. Wheelus and X. Zhu, "IoT network security: Threats, risks, and a data-driven defense framework," vol. 1, no. 2, pp. 259–285, 2020.

84. S. M. Karunarathne, M. Dray, L. Popov, M. Butler, C. Pennington, and C. Angelopoulos, "A technological framework for data-driven IoT systems: Application on landslide monitoring," vol. 154, pp. 298–312, 2020.

85. D. Sivaganesan, "A Data driven trust mechanism based on blockchain in IoT sensor networks for detection and mitigation of attacks," vol. 3, no. 01, pp. 59–69, 2021.

86. S. Rathore, B. W. Kwon, J. Park "BlockSecIoTNet: Blockchain-based decentralized security architecture for IoT network," vol. 143, pp. 167–177, 2019.

87. M. Li, W. Lou, and K. Ren, "Data security and privacy in wireless body area networks," IEEE Wireless Communications, vol. 17, no. 1, pp. 51–58, 2010.

88. S. Tan, D. De, W.-Z. Song, J. Yang, S. Das *et al.*, "Survey of security advances in smart grid: A data driven approach," IEEE Communications Surveys & Tutorials , vol. 19, no. 1, pp. 397–422, 2016.

89. S. T. Zargar, H. Takabi, and J. B. Joshi, "DCDIDP: A distributed, collaborative, and data-driven intrusion detection and prevention framework for cloud computing environments," in *7th International Conference on Collaborative Computing: Networking, Applications and Worksharing (CollaborateCom)*, 2011, pp. 332–341: IEEE.

90. F. Thams, A. Venzke, R. Eriksson, and S. Chatzivasileiadis, "Efficient database generation for data-driven security assessment of power systems," IEEE Transactions on Power Systems , vol. 35, no. 1, pp. 30–41, 2019.

91. D. Halperin, T. S. Heydt-Benjamin, K. Fu, T. Kohno, and W. Maisel, "Security and privacy for implantable medical devices," IEEE Pervasive Computing , vol. 7, no. 1, pp. 30–39, 2008.

92. K. Lorincz *et al.*, "Sensor networks for emergency response: Challenges and opportunities," IEEE Pervasive Computing , vol. 3, no. 4, pp. 16–23, 2004.

93. S. Chessa and P. Maestrini, "Dependable and secure data storage and retrieval in mobile, wireless networks," in *Proceedings of the International Conference on Dependable Systems and Networks*, 2003, pp. 207–216.

94. Q. Wang, K. Ren, S. Yu, and W. Lou, "Dependable and secure sensor data storage with dynamic integrity assurance," vol. 8, no. 1, pp. 1–24, 2011.

95. R. Di Pietro, L. V. Mancini, C. Soriente, A. Spognardi, and G. Tsudik, "Catch me (if you can): Data survival in unattended sensor networks," in *2008 Sixth Annual IEEE International Conference on Pervasive Computing and Communications (PerCom)*, 2008, pp. 185–194: IEEE.

96. J. Jacobs and B. Rudis, *Data-driven Security: Analysis, Visualization and Dashboards*. John Wiley & Sons, 2014.

97. S. Scheider, E. Nyamsuren, H. Kruiger, and H. Xu, "Why geographic data science is not a science," vol. 14, no. 11, p. e12537, 2020.

98. Y. Reddy, "Big data security in cloud environment," in *2018 IEEE 4th International Conference on Big Data Security on Cloud (BigDataSecurity), IEEE International Conference on High Performance and Smart Computing,(HPSC) and IEEE International Conference on Intelligent Data and Security (IDS)*, 2018, pp. 100–106: IEEE.

99. M. Abdelgadir, R. A. Saeed, and A. Babiker, "Cross layer design approach for efficient data delivery based on IEEE 802.11 P in vehicular ad-hoc networks (VANETS) for city scenarios," vol. 8, no. 4, pp. 01–12, 2018.

100. M. A. Elmubark, R. A. Saeed, M. Elshaikh, and R. A. Mokhtar, "Fast and secure generating and exchanging a symmetric keys with different key size in TVWS," in *International Conference on Computing, Control, Networking, Electronics and Embedded Systems Engineering (ICCNEEE)*, 2015, pp. 114–117: IEEE.

101. B. Jansen, J. Salminen, S.-G. Jung, and K. Guan, "Data-driven personas," Synthesis Lectures on Human-Centered Informatics (SLHCI), vol. 14, no. 1, pp. i–317, 2021.

102. R. Chi, Z. Hou, B. Huang, and S. Jin, "A unified data-driven design framework of optimality-based generalized iterative learning control," Computers & Chemical Engineering, vol. 77, pp. 10–23, 2015.

103. Y. Reddy, "Big Data Processing and Access Controls in Cloud Environment," in *2018 IEEE 4th International Conference on Big Data Security on Cloud (BigDataSecurity), IEEE International Conference on High Performance and Smart Computing,(HPSC) and IEEE International Conference on Intelligent Data and Security (IDS)*, 2018, pp. 25–33: IEEE.

104. R. Mokhtar, and R. Saeed, "Conservation of mobile data and usability constraints," *Cyber Security Standards, Practices and Industrial Applications: Systems and Methodologies*, edited by Junaid Ahmed Zubairi and Athar Mahboob, IGI Global, 2012, pp. 40–55.

105. R. Coulter, Q. L. Han, L. Pan, J. Zhang, and Y. Xiang, "Data-driven cyber security in perspective—Intelligent traffic analysis," IEEE Transactions on Cybernetics, vol. 50, no. 7, pp. 3081–3093, 2019.

106. Q. Liu, P. Li, W. Zhao, W. Cai, S. Yu, and V. Leung, "A survey on security threats and defensive techniques of machine learning: A data driven view," IEEE Access, vol. 6, pp. 12103–12117, 2018.

107. S. Karagiannopoulos, P. Aristidou, and G. Hug, "Data-driven local control design for active distribution grids using off-line optimal power flow and machine learning techniques," IEEE Transactions on Smart Grid, vol. 10, no. 6, pp. 6461–6471, 2019.

108. L. Li, S. X. Ding, and X. Peng, "Distributed data-driven optimal fault detection for large-scale systems," Journal of Process Control, Elsevier, vol. 96, pp. 94–103, 2020.

7 Data-Driven Techniques for Intrusion Detection in Wireless Networks

Lina Elmoiz Alatabani
Department of Data Communication and Network Engineering, The Future University, Sudan

Elmustafa Sayed Ali
Department of Electrical Engineering, Red Sea University, Sudan

Rashid A. Saeed
Computer Engineering Department, Taif University, Saudi Arabia

CONTENTS

DOI: 10.1201/9781003216971-9

7.1   INTRODUCTION

Data-driven approaches refer to the use of available data within a network for analysis and detection of unwanted traffic and to distinguish normal behavior from attack behavior. Using machine learning (ML) techniques, the performance of intrusion detection systems (IDS) has improved significantly to cope with the increasing focus on network quality, which is marked by a network that is secure, efficient, and reliable [1]. With the growing deployment of wireless networks today and with their future growth in applications that exchange important user information, such as financial information, the need for systems to detect unusual behavior in the network has increased. To address the importance of reliable IDS, models have been developed using ML techniques, which are known for their characteristics in feature selection, anomaly detection, pattern recognition, and outlier detection that increased systems efficiency and reliability in addressing security-related challenges [2]. For the importance of intrusion detection and with the known characteristics of ML techniques, this chapter provides details in data-driven intrusion detection systems.

The chapter is organized as follows: Section 7.2 provides brief concepts about wireless network attacks. The description of IDS and related data-driven techniques are reviewed in Section 7.3. In addition, different ML approaches used in data-driven IDS models with possible algorithms are discussed in Section 7.4. In Section 7.5, current models, challenges, and directions that guide future research efforts are discussed, considering data-driven applications of an IDS under wireless networks. Finally, the conclusion of the chapter is reviewed in Section 7.6.

7.2   WIRELESS NETWORK ATTACKS BACKGROUND

Modern wireless networks face a tremendous number of attacks because of important data circulating in the network. Attackers try to take advantage of the data for many purposes, including fraud, identity theft, and ransom [3]. There are several types of attacks, all of which lead to degrading the performance of the network and threatening the availability of services. Wireless networks are most vulnerable to attacks due to their computation and power limitations. Several levels of security are available to act as a reference for security practices and propose solutions accordingly. Level I is general security, which is based on the wireless equivalent privacy (WEP) algorithm that encrypts the data being transmitted. Level II represents the security threats not covered by Level I and are known as Wi-Fi protected access (WPA). Levels I and II cannot respond to all threats, which leads us to the highest level of security, Level III. Most of the weaknesses of the previous levels are covered, but this level of security is not used in all industries because of the degradation in performance due to the continuous running of scripts whenever a user attempts to perform a transaction.

Wireless mesh networks (WMN) are communication networks made of radio nodes organized in a mesh topology that seek to solve the limitations and improve overall network performance, which offers better security systems design. However, there are also several types of attacks facing WMNs, which leads researchers to develop systems and approaches to overcome these attacks [2, 3]. The most common threats breaching security levels are as follows.

a. ***Impersonation Attacks:*** This attack poses a great threat to WMNs because of the lack of proper authentication, allowing infected nodes to connect to the network and alter the routing information. An infected node might gain access to the network management system and may alter the configuration of the system [1].

b. ***Eavesdropping Attacks:*** In this attack, the attackers listen to ongoing communications between nodes to harvest connection information such as MAC address and cryptography.

c. ***Denial of Service on Sensing Attack (DoSS):*** The targeting of a physical layer application in a network environment where the attacker alters the data before it is read by the destination node. This leads to a false reading of the data and, therefore, a false decision [2].

d. ***Distributed Denial of Service Attack (DDoS):*** This type of attack targets a resource such as a network or a Web server with a large number of fake requests using botnets, which overwhelm the network or server with traffic beyond its capacity, causing a network crash and disruption in its usual service. A botnet is a term for a collection of compromised computers or "zombies" that are controlled by an external computer program, or bot herder. Botnets can be used for many purposes, including sending spam, phishing, and stealing information [3].

e. ***Node Capture Attack:*** A node capture attack is a type of attack technique that uses anomalous network behavior to redirect a target's network traffic to the attacker's computer, where it can be captured and examined. Hackers analyze their targets for vulnerabilities and access points typically by using node capture attacks. The term "node capture" generally refers to an attack on any software or hardware component that supports the formation of packets (e.g., routers, switches, firewalls) or that forwards packets (e.g., proxy servers) [4].

f. ***Selective Forwarding:*** This attack is responsible for the dropping of certain messages instead of delivering them whole. Infected nodes drop some of the messages and neighboring nodes are tricked into thinking that they are on a shorter route because of the dropped message gives the impression that the latency is reduced. By dropping more messages and forwarding less, selective forwarding preserves its energy level and, therefore, remains powerful to deceive the neighboring nodes [1].

g. ***Black hole Attack:*** A malicious node advertises itself using the routing protocol as the node with the shortest path to the targeted node to intercept its packet. Acting as a black hole, the attacking node drops all data packets passing by the node and the black hole node separates the network into two parts [5].

A mobile ad-hoc network (MANET) is a network with no fixed infrastructure, such as base stations, which means there are no centralized network administration services. A MANET is a self-configuring network of mobile nodes. Attacks on these types of networks are divided into two types as follows.

a. ***Data Traffic Attacks:*** This attack concentrates on using packets as the main attack. Packets pass by the nodes that form the network and some attacks aim at dropping packets while others aim at forwarding packets.

This eventually leads to degrading network quality and loss of important data [6].

b. *Control Traffic Attacks:* Making use of the existing routing protocols in the network, the decentralized nature of this type of network poses a great threat. Control traffic attacks use the routing protocol provided in MANETs (as there is no constraint on joining the network) and attackers create malicious nodes to join the network and seize the routing tables. These attacks can also use eavesdropping by taking control of the insecure routing protocol [6].

In the Global System for Mobile communication (GSM) hinking about cellular networks, people almost forget the issues of security as they unseeingly trust the Global System for Mobile communication (GSM) network. GSM has its weaknesses when it comes to security because cellular communications are complex in a number of areas, such as hardware, software, and configuration that are interconnected to deliver services such as voice, Internet, and messaging [5, 6]. The GSM has a standard that ensures the integrity and security of the network that is based on authentication, data confidentiality, and ciphering algorithms. However, there are still threats of attackers gaining control of a GSM network not related to network weakness of such as mobile phone malware, identity theft, and phishing [7]. The types of attacks facing the GSM network are stated below:

a. *Passive Attacks:* This attack usually observes, detects, or exploits the information in a certain system for a given purpose. However, the detected information remains the same without any change. Passive attack does not have an impact on the resources. Usually, this type of attack goes unnoticed because the purpose is to gain information about the system or to scan network ports for available vulnerabilities. Eavesdropping is one type of passive attack [8, 10].

b. *Active Attacks:* Unlike passive attacks, active attacks target the design vulnerabilities of the telecommunication infrastructure. These attacks begin by introducing a fake mobile base station that is under the attacker's control. The crucial vulnerability of using the fake base station is that the GSM does not require authentication of the network to the mobile phone [11]. This type of attack is called the International Mobile Subscriber Identity (IMSI) catcher in which the attacker gains control over communication parameters and other information such as encryption algorithms or traffic information [12]. The attack happens in between the victim's mobile phone and real base stations and is a type of Man-In-The-Middle (MITM) attack.

7.3   DATA-DRIVEN TECHNIQUES IN INTRUSION DETECTION SYSTEMS (IDS)

Network attacks are harmful actions aiming to degrade the overall network quality by targeting valuable information and services by exposing the system's confidentiality, integrity, and availability [13]. Cyberattacks have a significant role in losing a lot of money along with the negative impact to the economy if left without prevention.

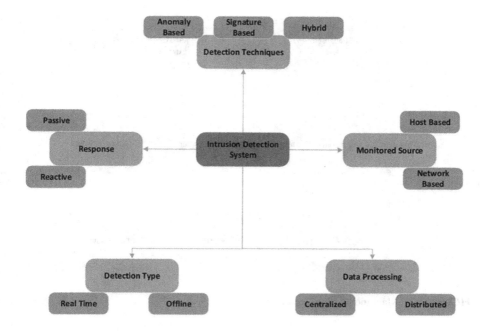

FIGURE 7.1   Intrusion detection system sections.

Although the exposure of modern wireless networks is computationally difficult for attackers, it is still not impossible in today's cyberspace [14, 15, and 16]. Therefore, the availability of an IDS is vital for attack prevention and to maintain the desired quality of service (QoS). There are two main categories in IDS (see Figure 7.1). The signature-based intrusion detection relies on the available databases of attacks and gives a trigger whenever a matching attack occurs. In addition, anomaly-based intrusion detection relies on the behavior of the network and uses the system characteristics to define the normal behavior of the network and detect any deviation from this normal pattern, and then gives a trigger when the deviation occurs [17,18].

The data-driven aspect in IDS is concerned with the use of ML. Although the two fields are remarkably close and may use combined techniques, we might see some slight difference in the operation of each concept, as these techniques' core concept is using data to reach a certain goal [19]. The newly proposed IDS using ML are often evaluated using certain datasets such as the KDD Cup 1999 dataset and University of New Mexico UMN dataset [20]. However, these datasets are no longer relevant for the modern systems of today because of the continual advances in computer technology. Using more recent datasets, such as Aegean Wi-Fi Intrusion Dataset (AWID), will have a better impact in today's research [21].

7.3.1 MACHINE LEARNING TECHNIQUES

There are many approaches involving the use of ML in intrusion detection categorized as follows: statistical methods, knowledge-based pattern matching, and expert

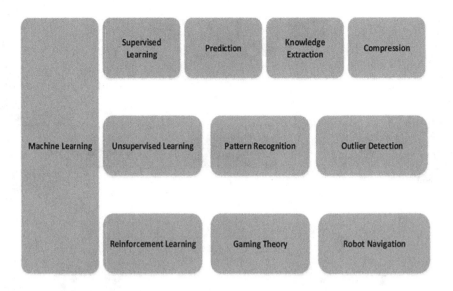

FIGURE 7.2   Machine learning techniques.

systems. In addition, there are also state transition-based methods, protocol-based analysis methods, soft computing methods, and combined methods (joining two or more methods) [22, 23]. Based on the approach, an ML technique is proposed to solve the intrusion detection problem in a manner that is efficient to cope with the rapid evolution of network systems [24]. ML techniques are divided into two main sections as shown in Figure 7.2. First are techniques based on artificial intelligence (AI), and second are techniques based on computational intelligence (CI).

AI techniques describe the methods that use classical AI, such as statistical modeling, and employ the two main branches of learning techniques, which are supervised and unsupervised learning. Researchers have proposed models to detect intrusions to networks using k-Nearest Neighbor (kNN) and multi-layer perceptron (MLP) for the supervised learning (SL) techniques and single-linkage clustering, k-means clustering, and γ-algorithm for unsupervised learning (UL) [25]. The UL showed proven results in detected unknown attacks while the SL techniques showed better performance in detecting known attacks; however, if there were unknown attacks hidden in the training data, the performance of SL algorithms decreased significantly [25, 26]. CI refers to the methods inspired by nature that solve complex issues that the classical approach is unable to solve, such as evolutionary computing, artificial neural networks (ANNs), and artificial immune systems. ML has been proposed for attack prediction rather than having a user predefined list of attacks in the network security domain. The use of ML hierarchical prediction for the prediction purpose leads to better-performing intrusion detection systems. The prediction models predict the exact type of attack and attribute it, i.e. DoS—(back, land, Neptune, pod, smurf, and a teardrop) [89].

a. **Genetic Algorithms (GA)**

Aiming at finding the best solution to a problem, a genetic algorithm (GA) presents each solution as a sequence of bits or (genes) with the name of the genome or chromosome mimicking the genetic system of a human body. The algorithm starts with a (population), i.e. set of genomes, and evaluates the population with a function that tests each genome for quality. The operation of the algorithm is stated as the use of reproduction operators called *crossover* and *mutation* preceded by *initialization* and *selection* to develop new solutions (descendants) which are evaluated for their performance in solving a given problem. While the mutation is a voluntary variation of a single gene, the crossover defines how the inherited properties in a population are defined from the parents to the decedents [25]. Several features characterize GAs in the way they perform a certain task: a) they can scan for solutions in a certain space in multiple directions at the same time, b) they are suitable for solving problems where the range of solutions is exceptionally large, c) they can be easily retrained to adapt to the generation of innovative solutions for intrusion detection, d) there is no need for complex mathematical operations, and e) they work with a range of different solutions rather than having to use one solution [26].

b. **Artificial Neural Networks (ANN)**

Resembling the operation of a human brain, ANNs consist of a collection of nodes (neurons) connected to form a network of nodes operating together to learn attack behaviors (see Figure 7.3). This type of network can learn by example using bounded, insufficient, and noisy data [27]. The main goal is to discover anomalies to describe user behavior with the use of a support vector machine (SVM) for SL in a high-dimensional environment. Consisting of three main components, ANNs have a layered feature

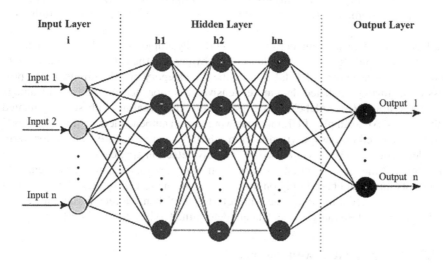

FIGURE 7.3   Artificial neural network architecture.

comprised of input nodes, hidden nodes for the processing of data coming from the previous layer, and an output node. There are several algorithms used in this type of networks: Multi-layered perceptron's backpropagation algorithm, radial basis functions based, neural tree, adaptive resonance theory-based, and Hopfield's networks [28].

c. **Artificial Immune Systems (AIS)**

Just like the human immune system fighting infections, the same concept is used to form a complete system consisting of different components to fight against attacks in a network, distinguishing normal from malicious activities. Researchers introduced six features based on immune systems that describe an effective intrusion detection system: distributed, self-organized, multi-layered, diverse, disposable, and lightweight [29].

d. **Deep Learning (DL)**

Deep Learning (DL) is the approach of extracting meaningful information from raw input data, which is often referred to as feature extraction of feature engineering. DL methods can learn to extract features independently from raw data without relying on customized features by experts, which gives a great advantage when dealing with unlabeled raw data. The advantage of feature extraction introduced by DL is to transform raw input data into features that accurately describe the problem and therefore increase the accuracy of the model [81, 82]. DL elevates the performance of existing network-based IDS (NIDS). The NIDS use network parameters such as packet length and flow size to extract certain features that could help with intrusion detection purposes, but the traditional NIDS suffer from high false-positive and false-negative rates, indicating that this IDS could fail in real-life applications. Host-based IDS are installed locally in servers and operate at the host level to monitor and analyze traffic details on the system level, meaning that it analyses the information related to system files, operating system, and system calls. These types of data are called audit trails.

DL has achieved outstanding performance in IDS by employing its abilities in features extraction and hidden sequential relationships by passing the network parameters into several hidden layers [83]. DL falls under the UL category of ML subsets, as the process of acquiring unlabeled data are expensive; it is more feasible to take advantage of a large amount of unlabeled data, especially when dealing with security-related applications. An example of UL methods would be an autoencoders, which is an ANN algorithms based on unsupervised learning [84]. Autoencoders are suitable for intrusion detection because they are used in anomaly detection applications, which rely on thresholds of the reconstruction error. In the training phase, the autoencoder learns to rebuild its input data based only on normal data. In the test phase, test data are input to the learned model to output the rebuilt test data. When the reconstruction error is higher than a certain margin, the data are determined to be abnormal [85].

7.3.2 GENERAL ML CLASSIFICATIONS USED IN IDS

SL is the process of mapping an input to an output based on a previously identified example. It also uses labeled training data, which is a designated label added to data, tagging them with a certain label, characteristic, or property. Different SL methods have been employed to serve IDS such as logistic regression (LG), SVM, and kNN [86]. UL represents the ML approach dealing with unlabeled data. Reinforcement learning (RL) is suitable for highly uncertain and complex environments. RL allows machines to learn to make sequences of decisions. The RL algorithm provides a range of characteristics, which allows the application of online IDS to have a rapid response. The advantages of the algorithms are represented as follows: It has a flexible reward function; after the training is finished, the policy function is a simple and quick neural network; the resulting trained neural network is suited to improve the process of distributed computing environments and offers a high attack prediction rate [87]. Applying the suitable RL algorithms depends on the identified problem and its requirements and assumptions. The available algorithms are actor-critic (AC), policy gradient (PG), deep Q-network (DQN), and double deep Q-network (DDQN) [88].

7.4 DATA-DRIVEN IDS MODELS AND ALGORITHMS

Models and algorithms have been developed with the influence of data-driven approaches to solve the issues facing the intrusion detection aspect of the networks. It has been an important topic, drawing researchers to investigate the different tools of ML to improve the security and efficiency of wireless networks.

7.4.1 DECISION TREE-BASED INTRUSION DETECTION

Based on two categories of classification, models have been developed on the basis of feature selection and classification. Using the dataset NSL-KDD, feature selection was implemented using two methods. The first is the correlation-based feature selection (CFS) and the second, is the consistency-based feature selection (CONS) using decision tree algorithms to classify network traffic as normal and abnormal, distinguishing attack behavior from normal behavior [30]. The classification under the decision tree model used a variety of decision tree algorithms, such as random forest and C4.5 under the KDD'99 dataset, and showed promising results when their performance was evaluated using precision, accuracy, and F-score having a high detection rate and low false alarm rate [31].

In the decision tree algorithm, the algorithm functions by calculating entropy, information gain, gain ratio, and finally, the split value, which concludes when the tree stops, splitting (see Figure 7.4) [32]. A leaf node is created based on the consideration of all examples belonging to the same class. Entropy is a metric that defines the impurity or uncertainty in observations, it is an information theory that decides how a decision tree selects to split data to leaf nodes. It can be calculated as:

$$Ent(S) = -\sum_{j=1}^{num_{class}} \frac{freq\,(Lj,S)}{|S|} * log2(\frac{freq\,(Lj,S)}{|S|}) \qquad (7.1)$$

where, Lj = a set of classes in use (L1, L2, ..., Lj), numclass represents the range of classes. In this algorithm, only two representations (normal and the anomaly or normal and attack) represent network traffic classification. S is the set of examples in

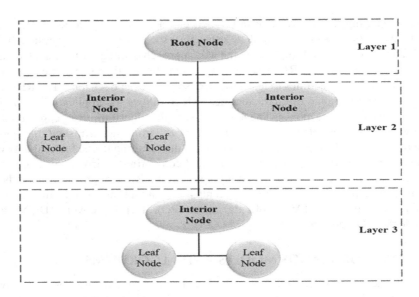

FIGURE 7.4   Decision tree model.

the training dataset [32]. Calculating information gain, which is the reduction in the entropy, by differentiating the entropy value of a dataset before and after variation

$$IG(a) = Ent(S) - \sum_{a_{val} \in \; values(a)} \frac{|S_a|}{|a|} * Ent(S_a) \qquad (7.2)$$

where S_a is a subset of S, a is the attribute, values (a) represent all possible values of attribute a and |a| is the sum of all values in attribute a.

To calculate the gain ratio, which is defined as reducing the bias, introduced by the information gain, it takes the number and size of decision tree branches when deciding which attribute to choose.

$$Gain \; Ratio(a) = \frac{IG(a)}{Split \; (a)} \qquad (7.3)$$

The Split value of an attribute

$$Split \; (a) = \frac{\left(\sum_{i=1}^{m} (a_{val})i \right)}{m} \qquad (7.4)$$

Where m = the number of values in attribute "a." After finding the attribute with the highest gain ratio, develop a decision node to divide the dataset in the given attribute and construct those nodes as decedents from the parent node. To visualize the performance of the algorithm, a confusion matrix is developed with four metrics [33]: The confusion matrix predicted classes corresponding to the four metrics are illustrated in table 7.1.

TABLE 7.1
Confusion Matrix

	Predicted Class +ve	Predicted Class –ve
Actual Class +ve	TP	FP
Actual Class –ve	FN	TN

- *True Positive (TP):* Represents the number of correct predictions.
- *True Negative (TN):* Represents the number of incorrect predictions for instances that belong to other classes.
- *False Positive (FP):* Represents the number of incorrect predictions within the same class.
- *False Negative (FN):* Represents the number of correct predictions within other classes.

Measuring the accuracy of a decision tree is vital for checking the percentages of correctly classified instances; the measurement is based on the confusion matrix [33]. The confusion matrix predicted classes corresponding to the four metrics are illustrated in Table 7.1 in the above paragraph.

$$Accuracy = \frac{TP + FN}{TP + FP + TN + FN} \tag{7.5}$$

Using a decision tree algorithm has impacted data-driven IDS significantly by improving the accuracy of intrusion detection. Depending on the dataset used in the training algorithm, such as KDD and Kyoto 2006+, which is a more recent dataset, the decision tree algorithm got a high true positive rate of 99% for distinguishing normal and attack traffic.

7.4.2 SUPPORT VECTOR MACHINE (SVM)

An SVM classifier is defined as the mechanism of dividing data into distinct categories by multi-dimensional hyperplane (N-dimension) that learns on a given dataset. It is based on statistical modeling theories. The instance of a training dataset is marked as $[(X_i, Y_i)]$, where $i = 1,2, \ldots, N$. N is the number of instances, X_i is the training dataset, and Y_i is the class of instances. The goal of SVM is to calculate the maximum margin that is dividing the hyperplane from the nearest point in a high-dimensional environment, meaning that SVM calculates the sum of the distance between a hyperplane and the nearest points of a dimensional space [34]. The boundary function can be represented as:

$$Minimize\ W(\alpha) = \frac{1}{2} \sum_{i=1}^{N} \sum_{j=1}^{N} y_i\ y_j\ \alpha_i\ \alpha_j(x_i, x_j) - \sum_{i=1}^{N} \alpha_i \tag{7.6}$$

With relation to:

$$\forall_i : 0 \leq \alpha_i \leq C, \text{ and } \sum_{i=1}^{N} \alpha_i y_i = 0, \tag{7.7}$$

where C is the soft margin parameter, $C > 0$ and α is a vector of several variables N. Kernel functions can be used by SVMs to divide the instance of data into subcategories. These functions are represented as:

$$\text{Linear kernel}: k(x_1, x_j) = x_i^T . x_j \tag{7.8}$$

$$\text{Polynomial kernel}: k(I, x_j) = \left(\gamma x_i^T . x_j + r\right)^d, \gamma > 0 \tag{7.9}$$

$$\text{Radial bases function kernel}: k(I_i, x_j) = \exp\left(-\gamma \|x_i - x_j\|^2\right), \gamma > 0 \tag{7.10}$$

$$\text{Sigmoid kernel}: k(I, x_j) = \tanh\left(\gamma x_i^T . x_j + r\right) \tag{7.11}$$

Where γ, r, and d represent the kernel parameters.

SVM is an algorithm that classifies data by a hyperplane or a line while the hyperplanes are linear, thus, it can be represented mathematically. The most effective hyperplane is the hyperplane with the maximum margin [35]. SVM is one of the SL models that has high performance in pattern recognition and classification with data that is high dimensional (see Figure 7.5) [36]. It is also known to be suitable with applications such as intrusion detection with its proven results of low generalization error and high detection rate [37]. SVM divides the dataset into two parts: training and testing sets. It builds a model using the training dataset that predicts the target values of the test dataset based on its attributes [38].

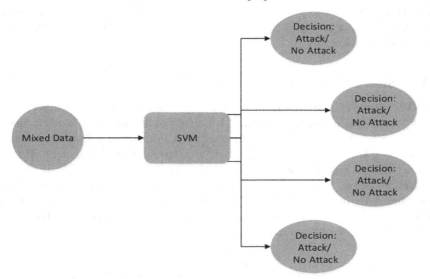

FIGURE 7.5   Support vector machine mechanism.

IDSs use different SVM techniques such as quadratic SVM, fine Gaussian SVM, linear SVM, and medium Gaussian SVM. Linear SVM uses the linear kernel to separate classes, which gives 96.1% accuracy in detecting abnormal network traffic [39]. Quadratic SVM uses a quadratic kernel and gives an accuracy of 98.6% for intrusion detection. Fine Gaussian SVM develops fine complete differences between classes and gives 98.7% accuracy. Medium Gaussian SVM uses a Gaussian kernel to create fewer differences between classes where it gives an accuracy level of 98.7%. SVM techniques are used for data-driven IDSs in the context of training, decision making, and testing using datasets that consist of current traffic data from networks where it showed a significant impact on the accuracy of detecting malicious traffic.

7.4.3 NAÏVE BAYES

Naïve Bayes a classifier, which is based on the original Bayes' theorem. With powerful (naïve) independence assumption, it eliminates the assumption of the existence of a correlation between features [40]. The classifier can be trained and performs well in a SL environment with the consideration that the existence or nonexistence of certain features in a class is not related to the existence or nonexistence of any other feature [41]. The overall algorithm consists of two phases, learning and prediction. The classifier operates to label the classes by performing calculations to instances x in a dataset X that belongs to one class C [42]. Using unconditional probability with the consideration of posterior values (the conditional probability of x) in class Ci, given the simplified Bayesian analysis with the equation:

$$P(C = C_i | X = x) = \frac{P(C_i) \prod_{j=1}^{d} P(x_l | C_i)}{P(x)} \qquad (7.12)$$

Naïve Bayes uses the maximum likelihood method instead of Bayesian probability, assuming the probability density function (PDF) to calculate $P(x_l | C_i)$

$$P(x_l | C_i) = \frac{1}{\sqrt{2\pi\sigma_i^2}} e^{-\frac{x_j - \mu_i^2}{2\sigma_i^2}} \qquad (7.13)$$

The naïve Bayes classification is aimed at updating the average and standard deviation σ of the features gradually. The training phase uses the first N number of connection samples Sn, which is calculated by acquiring the parameters mean μ for all continuous features. Then it recalculates the mean and standard deviation of class C to update the μ and σ values after each test sample using simple and exponential equations. The values of μ and σ for each feature are updated when a new sample v is available [43].

$$\mu_i' = \frac{k_i \mu_i + 1}{k_i + 1} \qquad (7.14)$$

$$\sigma_i' = \sqrt{\frac{(k_i + 1)(x_j^2 + k_i(\sigma_i + \mu_i)) - (x_j + k_i \mu_i)^2}{(k_i + 1)^2}} \qquad (7.15)$$

Where k_i represents the number of samples of class i, μ_i' and σ_i' are the updated values of the mean and standard deviation. x is the instance in the dataset. Weighted multiplier α can be used to obtain the recent values of the mean and standard deviation by considering their previous values exponentially by:

$$\mu_i' = (1-\alpha)x_j + \alpha\mu_i \qquad (7.16)$$

$$\sigma_i'\sqrt{(1-\alpha)\sigma_i^2 + \alpha(x_j - \mu_i)^2} \qquad (7.17)$$

where α is to determine the learning rate of the naïve Bayes classifier. The naïve Bayes classifier addressed the problems in the anomaly-based intrusion detection approaches where they suffer from false alarm rates due to their struggle in building an adaptive model, and the issue of concept drift where the observed entity behavior changes due to differences in a task, time, and other uncertainty constraints [44].

With the implementation of online naïve Bayes classifiers in intrusion detection datasets and the use of exponential weights to calculate the average and standard deviations, the idea of giving more weights to the most current samples allows naïve Bayes to adapt to the recent attacks developed over time. Naïve Bayes is efficient in applying online IDS with regard to the time and accuracy of distinguishing normal and abnormal network traffic. The naïve Bayes classifier is a statistical anomaly-based IDS that is useful in detecting unknown attacks, unlike the signature-based IDS, which has advantages in detecting known attacks because it trains on a dataset containing the available known attacks. Naïve Bayes dynamically detects an anomaly in network traffic to differentiate anomalies that diverge from normal behavior.

7.4.4 REGRESSION

Most suitable for SL approaches, logistic regression is most useful for binary responses, but with the efforts of continuous enhancement of learning algorithms, the operation can be extended to cover multi-class purposes [45]. Using conditional likelihood to determine the logistic loss function with the consideration of conditional distribution [46], logistic regression aims to develop a model that discovers a relationship between the class and features by predicting a value between 0 and 1 if only two classes are considered. [47]. As logistic regression operates within two values, 0 and 1, a sigmoid function is used to ensure that the output remains within the 0 and 1 range [48].

$$h_\theta(x) = g\left(\theta^T x\right) = \frac{1}{1+e^{-\theta^T x}} \qquad (7.18)$$

Where θ represents an unknown vector and x is the feature values vector. Assuming N number of features in the dataset

$$\theta_0 + \theta_1 x_1 + \cdots + \theta_i x_i \qquad (7.19)$$

where θ_0 is the bias parameter, θ_i are unknown parameters between 1 and N $(1 \le i \le N)$, and x_i are the values of the feature. The objective function is represented as:

$$Cost\left(h_\theta(x), y\right) = \begin{cases} -\log\left(h_\theta(x)\right) if \ y = 1 \\ -\log(1 - h_\theta(x)) if \ y = 0 \end{cases} \quad (7.20)$$

Where y is the class label considering its value as $y \in \{0,1\}$. The cost function can be further simplified to calculate the value of parameter θ as:

$$J(\theta) = \frac{1}{m} \sum_{i=1}^{m} Cost\left(h_\theta\left(x^{(i)}\right), y^{(i)}\right) \quad (7.21)$$

$$= -\frac{1}{m}\left[\sum_{i=1}^{m} y^{(i)} \log h_\theta\left(x^{(i)}\right) + \left(1 - y^{(i)}\right)\log(1 - h_\theta(x^i))\right] \quad (7.23)$$

where m is the number of samples and y is the class label. A gradient descent method is used because of the rounded nature of the objective function, represented by:

$$\theta_j := \theta_j + \alpha\left(y^{(i)} - h_\theta\left(x^{(i)}\right)\right) x_\theta(i) \quad (7.24)$$

where θ_j defines the unknown parameter value and α is harvested using trial and error depending on the class values. To give more accuracy to the objective function, regularization is proposed to improve the values obtained by logistic regression. The reason behind using regularization is to minimize additional feature's effect on the objective function, therefore, the coefficient values of features are decreased.

$$J(\theta) = -\frac{1}{m} \sum_{i=1}^{m}\left[y^{(i)} \log\left(h_\theta\left(x^{(i)}\right)\right) + \left(1 - \left(y^{(i)}\right)\right)\log\left(1 - h_\theta\left(x^{(i)}\right)\right)\right] + \frac{\lambda}{2m}\sum_{j=1}^{n}\theta_j^2 \quad (7.25)$$

where λ is the regularization parameter, which is added to the objective function to give more accuracy; therefore, the unknown values are captured using:

$$\theta_0 := \theta_0 + \alpha \times \frac{1}{m}\sum_{i=1}^{m} h_\theta\left(x^{(i)} - y^{(i)}\right)x_0(i) \quad (7.26)$$

$$\theta_j := \theta_j + \alpha \times \frac{1}{m}\sum_{i=1}^{m} h_\theta\left(x^{(i)} - y^{(i)}\right)x_j(i) - \frac{\lambda}{m} \times \theta_j \quad (7.27)$$

where θ_0 is the bias parameter and m is the number of samples [49].

Logistic regression as shown in Figure 7.6 is used as a classification approach in feature selection of the attack network traffic. It detects the most crucial features in each class used. The logistic regression classifier is linear, it is most useful in binary systems and for multi-class classification with an accuracy level of 91.5% for feature selection.

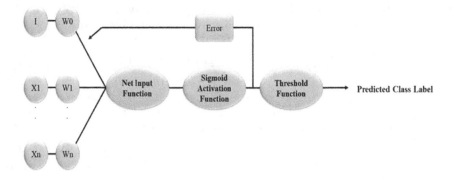

FIGURE 7.6   Logistic regression model.

7.4.5   CLUSTERING APPROACH

Clustering is a grouping methodology made to solve complex dimensional problems, clustering groups' features that are relevant to each other and eliminating irrelevant features. Cluster analysis observes values of features using the Pearson correlation coefficient and mutual information or using the most optimal feature. Feature selection, which is based on cluster analysis, is performed by discovering features after increasing their diversity; the grouping is done following three stages: developing a design structure of feature space, grouping features that are relevant, and selecting a representative feature for each cluster (see Figure 7.7) [50]. Several clustering algorithms serve the clustering approaches such as k-means clustering, naïve Bayes, and other hybrid algorithms.

IDS using cluster analysis involves selecting the applicable feature that might affect the classification result and using the Pearson correlation to calculate the

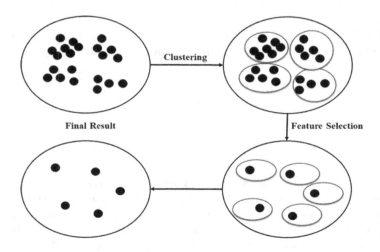

FIGURE 7.7   Clustering general architecture.

feature evaluation value, which is used to determine the relationship between variables. When there is a variation in one variable, it prompts the other to change. To have an IDS with high performance is evaluated by the quality of extracted data. The application of clustering and ML classifiers adds value to intrusion detection systems with the use of appropriate cluster formation approaches, which have a significant impact on the selected feature. Feature reduction improves the accuracy, sensitivity, and precision of IDSs.

a. **K-means Clustering Algorithm**

The simplest and most popular clustering algorithm groups features based on sample similarity. Considered as UL, k-means splits a cluster by decreasing the clustering error. The basic idea is to randomly select the initial center from the data object and then operate to determine the distance between each data object and each center [51]. The clustering adopts partitioning methodology where objects are arranged into k partitions and it spots outliers in the data. When the value is distant from most data, the mean value of a cluster will be remarkably deformed [52].

$$F(x_1, x_2, \ldots, x_N) = \sum_{k=1}^{m} \sum_{x_i \in c_k} \|x_i - \bar{x}_k\|^2 \tag{7.28}$$

Where: $x_1, x_2, \ldots, x_N$ is a set of instances, K is a predefined number of clusters by k-means algorithm, c_k represents the kth cluster.

$$\bar{x}_k = \frac{1}{n_k} \sum_{x_i \in c_k} x_i \tag{7.29}$$

$\bar{x}_k$ is the center of the kth cluster and n_k is the number of instances within the kth cluster, $\|.\|$ represents the Euclidean norm used by the algorithm, which measures the distance between data and the center of the cluster using the equation:

$$d(x, z) = \sqrt{\sum_{j=1}^{d} (x_j, y_j)^2} \tag{7.30}$$

where d represents the number of dimensions or features and x_j, y_j are the values of the attributes. This might have a limitation because not all attributes exist in the ideal environment, which has spherical homogeneity in a cluster. Therefore, the use of information gain and entropy is proposed using weighted Euclidean to overcome this issue [53]. The equation will be represented as:

$$d(x, z) = \sqrt{\sum_{j=1}^{d} w_j (x_j, y_j)^2} \tag{7.31}$$

where w_j represents the weight of the jth attribute.

K-means-based self-organizing map clustering algorithms can elevate the performance of intrusion detection systems, as it decreases the number of clustering iterations, which might affect the computation performance. As the number of clusters increases, the efficiency of the k-means algorithm increases, and should happen concerning the exact number of data types given. The k-means clustering algorithm is known for its faster clustering speed; thus, it has higher efficiency and higher performance in providing more trusted data to support a comprehensive network security approach.

b. **K-Nearest Neighbor (k-NN)**

Used in pattern recognition, the k-nearest neighbor is non-parametric and used for classification. It discovers the resemblance between unlabeled samples and training samples. It operates in a multidimensional feature space where each sample is given a class label in the classification phase of the model where k is a constant value defined by the user [54]. The k-NN algorithm pursues a spatial pattern for a training record that is near the k-training record (hence "k-Nearest Neighbor"). The closeness is defined by the Euclidean distance between two records X_1 and X_2 using the equation:

$$dist(X_1, X_2) = \sqrt{\sum_{i=1}^{n} (x_{1i} - x_{2i})^2} \qquad (7.32)$$

calculating the accumulative distance measure by taking the square root [55].

K-nearest neighbor is an instance-based learning algorithm that does not start learning until it's time to start making predictions, no training period or step is required, which makes it faster than another algorithm that learns first, then predicts. While it does not start the learning process until it's time for prediction, this gives a chance to add new data without negatively impacting the algorithm. This makes its implementation easy because we need only two parameters for the algorithm, which are the distance and the value of K.

Nonetheless, there are minor issues in implementing k-Nearest Neighbor in IDSs. These are represented in the performance issues found when dealing with large datasets and using them in high-dimensional environments, which leads to the introduction of hybrid approaches (combining K-NN with other learning algorithms) to have more efficient results.

7.4.6 Genetic Algorithm (GA)

A GA is a search method that is based on natural selection. It is a random search approach that starts with a set of individuals or chromosomes (the population) that consists of a group of genes that can be represented as bits. The population is selected based on its fitness, meaning that if the fitness value is higher than the selection, chances increase. The GA model has simple steps beginning with randomly selecting individuals from the sample space S, evaluating the fitness of the individuals in the population P, repeating the process of fitness evaluation until the condition is met, and finally selecting the most appropriate individual from the population [56]. The fitness function is represented by:

$$fitness(j) = \frac{TP_j}{\sum_{i \in population} TP_i} - FPR_j \tag{7.33}$$

$$FPR_j = \frac{FP_j}{FP_j + TN_j} \tag{7.34}$$

where TP_j represents true positive, FPR_j is the false positive rate, and TN_j is the true negative for the chromosome j

$$fitness(j) = \frac{(TP_j + TN_j)}{n} \tag{7.35}$$

where n is the total number of events. In a perfect environment, the number of normal events should be greater than the number of intrusions or attacks, therefore, the above equations should be altered to adapt to a more flexible scenario.

$$fitness(j) = \frac{TP_j + 1}{n} \; exp\left[-\left(t_1 \frac{FP_j}{n}\right)^2\right] - t_2 F \tag{7.36}$$

Where F represents the number of selected features, t_1 is a constant that corresponds to the intolerance of FP and is a factor used to sentence individuals with a considerable number of selected features.

Which method to use is a question that is addressed according to the type of attack facing the network and available features. Proposing n number of attacks, a subset can be chosen and represented by n_0 and can be statistically treated using binomial distribution considering the proportion of TP as the parameter of p. Given by the equation with a confidence level of 95% [57]:

$$\frac{TP}{TP+FP} \in \left[\hat{p} - 1.96\sqrt{\frac{\hat{p}(1-\hat{p})}{n_0}} - \varepsilon, \hat{p} - 1.96\sqrt{\frac{\hat{p}(1-\hat{p})}{n_0}} + \varepsilon\right] \tag{7.37}$$

$$\hat{p} = \frac{\widehat{TP}}{\widehat{TP} + \widehat{FP}} \tag{7.38}$$

The introduction of AI technologies has occurred to cope with a large amount of network data and to maximize IDS performance by minimizing the time and complexity of the classification model by selecting only the useful features that serve the purpose of intrusion detection. The use of the GA has contributed to selecting the minimum number of related features. The application of GA has helped in elevating the performance of the classifiers that are implemented in the classification phase along with improving the time taken for the learning process and reducing the complexity of the system. GA algorithms apply a repetitive approach to anomaly-based detection systems, which improves the performance of the IDS.

7.4.7 ARTIFICIAL NEURAL NETWORK (ANN)

Taking the features of the biological neural network, ANN bears resemblance to the operation of the human brain in the classifications of data, which makes decision-making more efficient. ANN consists of many neurons or nodes connected and operating in a parallel manner, which is useful in reducing time to process data [58]. The neurons operate to produce complex hypotheses and the complexity increases with the increase of the number of neurons. The operation of the input nodes is made in a feedback process to evaluate the hypotheses, and the flow of events is spread through the network until it reaches the output where the event is labeled as normal or attacked. The gradient of the cost function can be calculated when the gradient descents are introduced to send the error found in the output node back to the network using a backpropagation process [59, 60]. The neural network model undertakes training to learn the pattern created in the overall system by two learning procedures, supervised and unsupervised learning. In SL, the network is provided with a labeled training dataset. MLP is an example of ANN that is trained using SL where the UL uses unlabeled data. Self-organizing maps (SOMs) is an example of UL that constructs a low dimensional, decentralized representation of the input data called map [61].

a. **Multi-Layer Perceptron (MLP)**

The process of training is complicated in MLP; thus, it can be considered as an optimization problem, which has a substantial number of algorithms to address this problem. This type of learning algorithm has been proven a powerful mechanism to address challenges facing the available intrusion detection approaches. It gained importance for its characteristics in establishing user behaviors using statistical models; moreover, it has an extendable, open, and resilient structure. The simplest type of ANN consists of three layers: input layer, fixed hidden layer, and an output layer [62], which can be represented by:

$$y_i = f_i \left(\sum_{i=1}^{n} \omega_{i,j} . x_i + \beta_i \right) \tag{7.39}$$

Each artificial neuron or node receives an input signal, which is represented by x_i that is correlated with a weight $\omega_{i,j}$ to give power to the input signal. Where f_i is an activation function used to calculate the output signal concerning the input signal.

The MLP representation contains a series of weights and biases where the length of the solution vector is determined by the number of weights and biases, which also depends on the number of hidden layers and the nodes within a hidden layer. The length can be calculated using the following: *Length of weights and biases vector* $= W + B$.

$$W = (1 \times N) + ((N \times N) \times (H - 1)) + (N \times O) \tag{7.40}$$

$$B = H \times N + O \tag{7.41}$$

Where W is the number of weights, B is the number of biases, I is the number of nodes in the input layer, N is the number of nodes in each hidden layer, H is the number of the hidden layers, and O represents the number of nodes in the output layer [63].

The main objective of the MLP algorithm is to deliver the highest classification and prediction accuracy for both the training and testing phases. Thus, calculating the fitness function concerning the previous functions is feasible using the following:

$$f(S_i) = Sigmoid \ (s_j) = \cfrac{1}{\left(1 + \exp\left(-\left(\sum_{i=1}^{N} W_{ij}\Theta \ \chi_i - \beta_i\right)\right)\right)},$$
$$j = 1, 2, \ldots, H \tag{7.42}$$

Where n is the number of the input nodes. W_{ij} is the connection weight from the nodes in the input layer to the input nodes in the output layer. β_i is the bias of the jth hidden node, and χ_i is the ith input.

$$S_i = \sum_{i=1}^{n} W_{ij} \cdot \chi_i - \beta_i \tag{7.43}$$

Next, the cumulative output is calculated using:

$$O_k = \sum_{i=1}^{N} W_{kj} \cdot f(S_j) - \beta_k, k = 1, 2, \ldots, O \tag{7.44}$$

Where W_{kj} is the connection weight from the jth hidden node to the kth output node, and β_k is the bias of the kth output node. Next, we calculate the learning error represented by E

$$E_k = \sum_{i=1}^{O} \left(O_i^k - d_i^k\right)^2 \tag{7.45}$$

$$MSE = \sum_{k=1}^{q} \frac{E_k}{q} \tag{7.46}$$

where q is the number of training samples, d_i^k is the desired output of the input unit, O_i^k is the actual output of the input unit, thus, the fitness function can be represented as:

$$Fitness(x_i) = MSE(x_i) \tag{7.47}$$

MLP, as represented in Figure 7.8, contributed to the performance optimization of IDS by the concept of multiple layers of networks having input, hidden, and output layers of connected neurons. Useful for the classification phase of the intrusion detection system, MLP has given a high classification rate with reduced error rates although it requires time in the learning

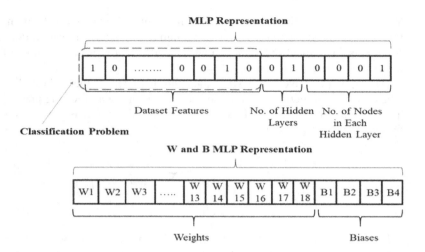

FIGURE 7.8 Multi-layer perceptron algorithm.

process. It is robust in adding new data to the algorithm, which does not require retaining the whole network to add data.

b. **Self-Organizing Maps (SOM)**

A SOM transforms a high dimensional data space into a low dimensional data space; it groups vectors that are similar and then displays them graphically [87 subs from SOM]. SOM has been implemented in areas like pattern recognition, anomaly detection, and attack detection, with the ability of the SOM network to distinguish normal data from abnormal data. SOM has two layers: the input layer, which accepts high dimensional data, and the output layer, which oversees analyzing and comparing input data, where each neuron in the layer constitutes to a class that needs to be clustered [64–66].

A competitive learning algorithm, which reviews the distribution of input samples. In the SOM model, the neuron of the input layer and output layer are fully connected, the output is represented via an array (two-dimensional array). The competitive learning algorithm calculates the Euclidean distance between the input patterns and the weights W of the output layer, where the shortest distance is the desired result known as Best Matching Unit (BMU) or "winning" neuron [67]. The formula is represented as:

$$D(X, C_k) = min\{\|X - W\|\} \tag{7.48}$$

where C_k is the winning neuron, and W is the weight vector of the neuron in the output layer. After producing the winning neuron k, the weight should be recalculated to have the updated values as:

$$\omega_k(new) = \omega_k(old) + \rho\varphi(k,n)(x - \omega_k) \tag{7.49}$$

where $\varphi(k,n)$ is the neighborhood function, that distance is minimized between k and n represented by $\|W_k - W_n\|$, neuron n is the neighbor of winning neuron k, and ρ is the learning rate, which decreases over time. The neighborhood function is represented by:

$$\varphi(k,n) = \begin{cases} e^{\left(-\|W_k - W_n\|^2\right)/2\sigma^2}, & k \neq n \\ 1, & k = n \end{cases} \tag{7.50}$$

where σ^2 is the width parameter that decreases over time. This concept is used to maximize the intrusion detection rate, but the calculation of all input data with every neuron in the output layer presented the challenge of large computational capacity. Therefore, some variations of the regular SOM algorithm were proposed, such as hierarchical SOM and growing hierarchical SOM (GHSOM).

The traditional SOM has some limitations in the computation overhead and online detection, thus, improvements of SOM have been introduced to overcome these limitations. GHSOM is capable of dynamically expanding horizontal and vertical neuron mapping, which leads to the reduction in computation load, achieving qualities of self-learning and self-adaptation.

The investigation of SOM feature selection abilities presents that while the number of features decreases, the detection rate does not lessen. Reducing the size of data leads to improvements in anomaly detection, computation cost, and the overall algorithm performance, which also leads to efficient IDS. The variations of SOM have proven promising in IDSs, which is represented in the high detection rates and adaptation to online detection. However, there is a need for improvements in other aspects, such as high false alarm, which leads to inefficient SOM mapping and high complexity in the classification calculation.

7.4.8 ARTIFICIAL IMMUNE SYSTEM (AIS)

Typically, based on the human immune system differentiating between good and bad cells, the Artificial Immune System (AISs) separates normal from attack behavior within a network. AIS is a computational model that has a range of algorithms, which are Clonal Selection, Negative Selection, Immune Network, Danger theory, and other hybrid algorithms that combine two or more of these algorithms. AIS is used in the intrusion detection field to detect intrusions in real time [68]. Designing AIS for intrusion detection consists of steps or phases, starting with representing the network traffic in the same way the immune system shows antibody or antigen encoding—with the representation of the self/non-self-manner. The representation should be based on feature elements and measuring affinity.

a. **Negative Selection Algorithm**

The negative selection algorithm (NSA) is an algorithm that revolves mainly around two major steps: the generation of the set of detectors, where each detector corresponds to a numerical range, and the monitoring of data for changes [69]. The non-self-value is represented by N-dimensional space and non-self-radius Rs using the equation [NSA EQ]:

$$S = \{X_i \mid i = 1, 2, \ldots, m; \ R_s = r\} \tag{7.51}$$

Where X_i are points in the normalized n-dimensional problem space.

$$X_i = \{X_{i1}, X_{i2}, X_{i3}, \ldots, X_{iN}\}, \ i = 1, 2, 3 \ldots, m \tag{7.52}$$

The space can be defined as S=1-NS where S is self, and NS is non-self

$$d_j = \left(C_j, R_j^d\right) \tag{7.53}$$

where d_j is the detector with center $C_j = \{C_{j1}, C_{j2}, C_{j3}, \ldots, C_{jN}\}$, while Rj is the detector's radius and Rd is the detector's diameter. The Euclidean distance equation is used to calculate the distance between non-self-sample X_i and detector d_j

$$L\left(X_i, d_j\right) = \sqrt{\left(X_{i1} - C_{j1}\right)^2 + \cdots + \left(X_{iN} - C_{jN}\right)^2} \tag{7.54}$$

where, $L(X_i, d_j)$ values are compared with the threshold Rs, acquiring the match value of $\ltimes$

$$\ltimes = L\left(X_i, d_j\right) - R_s \tag{7.55}$$

The detector threshold Rd, j of detector d_j can be represented as

$$R^d, j = min(\ltimes) \ if \ltimes \leq 0 \tag{7.56}$$

NSA is applied to monitor the changes within network activity to detect malicious activities. It was tested against accuracy and execution time and showed that its performance in classification decreases when increasing the number of instances. When it comes to the execution time, it takes longer to process when the sample size is large. Increasing the number of detectors in a sample improved the classification accuracy. Research indicates that NSA suffers from issues when dealing with large sample sizes, thus; improvements must be applied to have better results or apply NSA only to small-scale networks.

b. **Clonal Selection Algorithm**

This algorithm is based on the clonal selection theorem where detectors are multiplied based on the affinity. Detectors with low affinity are replaced by cloned detectors [68]. The algorithm consists of a series of steps: 1) generate and spread a set of detectors across the problem space, 2) calculate the affinity, 3) organize the detector based on their affinity where the lowest affinity is represented as the smallest difference, 4) clone the detector with the lowest affinity, 5) transform the detectors using the pre-defined ratio, 6) calculate the new affinity value, 7) leave the confident detectors and remove the less confident ones, 8) randomly generate new detectors to replace the deleted detectors, and 9) repeat the steps until the predetermined threshold is satisfied [70]. Traditional clonal selection algorithms have limited performance, suffering from low detection accuracy and high false-alarm rates, the introduction of the improved algorithm has helped in IDSs that cope with the growing use of wireless networks. The improvement aims to develop a considerable number of memory cells to apply them in the classification of test data. The improved clonal selection algorithm showed detection rates within 99% and a minimum error detection rate. The algorithm fits the application of IDS in a large amount of data, which makes it ideal for large-scale network applications.

7.4.9 AUTO ENCODERS

Autoencoders are used for anomaly detection to detect abnormal behavior within network traffic based on reconstruction error applied to the network-based IDS. There are several types of autoencoders applied depending on the requirements and characteristics of the system. To have the optimum performance of the algorithm, an environment with altering degrees of abnormal behavior should be taken under consideration when training the model. Autoencoder algorithms aim to extract features of input data by reconstructing the original input data. This is done in two stages, encoding and deconstruction.

a. **Encoding**

Using encoding parameter $\theta = \{W, b\}$, an input x is mapped to a hidden representation where W is the weight and b represents bias. The hidden representation is stated as follows,

$$y = f_\theta(x) = s(W_x + b) \tag{7.57}$$

b. **Mapping the Reconstruction Vector**

In this stage, the hidden representation y is then mapped to a reconstructing vector with the use of decoding parameters $\theta' = (W', b')$ represented by:

$$z = g_{\theta'}(y) = s(W'y + b') \tag{7.58}$$

The autoencoder model minimizes the average reconstruction error by mathematically representing the parameters with the following formulas:

$$\theta^*, \theta^{*'} = arg\min_{\theta,\theta'} \frac{1}{n} \sum_{i=1}^{n} L(x^{(i)}, z^{(i)}) \tag{7.59}$$

$$\theta^*, \theta^{*'} = arg\min_{\theta,\theta'} \frac{1}{n} \sum_{i=1}^{n} L(x^{(i)}, g_{\theta'}(f_{\theta(x^{(i)})})) \tag{7.60}$$

where L is an ordinary loss function, such as $L1(L(x,z) = |x - z|)$ or $L2(L(x,z) = |x - z|^2)$. The loss function can be substituted by a reconstruction cross-entropy function with the consideration of x and z being bit vectors or bit probability

$$L_{\mathbb{H}}(x,z) = \mathbb{H}(\mathbb{B}_x \| \mathbb{B}_z) \tag{7.61}$$

$$L_{\mathbb{H}}(x,z) = -\sum_{k=1}^{d} [x_k \log z_k + (1 - x_k) \log(1 - z_k)] \tag{7.62}$$

where x is a binary vector and $L_H(x, z)$ is a negative log probability for x concerning parameter z and k is the count until the kth parameter. To simplify the operation of the autoencoder, encode and decode the training data x by mapping it to a hidden representation y, which is mapped again using a reconstruction vector z [86].

7.5 DATA-DRIVEN BASED IDS CHALLENGES AND FUTURE DIRECTIONS

When designing an intrusion detection model, there are some factors to consider such as computational complexity (which leads to delay in the performance), the efficiency of the model, detection rate, and false-alarm rate [71, 72]. The challenges are represented through a series of issues facing currently proposed IDSs, the challenges are:

a. *High False-Alarm Rates:* Depending on the source, the training data might be overly complicated or might lack the required clarity, being redundant. This leads to incomplete training, which causes false-alarm rates to increase [72].

b. *Communication Overhead:* High communication overhead is often caused by the handling or transmission of IDS model parameters, which leads to server bottlenecks.

c. *Resource Management:* Depending on which type of network the IDS is operating, some network devices have low power, which leads to low computation capacity. Minimum high-communication support often leads to delays in the system.

d. *Vulnerable Intrusion Detection Setup:* Difficulty in separating new anomalous traffic in a diverse network environment.

e. *Accurate Deployment of Intrusion Detection Systems:* When deploying an IDS, there are often a vast number of parameters that need fine-tuning, thus, any difference might affect the desired results. Manually adjusting the parameters is a time-consuming process [72].

f. *Timely Risk Management:* Risks must be accounted for when designing IDSs to reduce attacks; choosing the optimum response maximizes the security performance against attackers on a network [73].

To address these challenges, future research in developing data-driven IDSs must focus on hierarchical clustering techniques and RL approaches to improve false-alarm rates [74]. Binaries neural network (BNN) is an approach to improve the performance and latency in traffic classification. BNN is a subtype of a deep neural network (DNN), which is a neural network with binary weights and activations. The BNN computes the parameter gradient to reduce the memory size of modern communication protocols in 5G and 6G networks, adding more reliable intrusion detection [75, 76]. It is an executing technique for optimization of resource allocation and lightweight ML and DL techniques to increase the overall throughput [77, 78]. It also utilizes the execution of blockchain to secure the communication between the server and clients [79] and applies the optimization algorithms to adjust the parameters [80].

7.6   CONCLUSION

Data-driven techniques for intrusion detection have gained significant importance in the deployment of wireless networks because of the increasing focus on network security to protect user privacy and data integrity. The research concentrated the effort toward a more reliable and efficient IDS by using today's state-of-the-art techniques of ML, the appropriate models and algorithms designed to detect unusual behavior within the network, and applying the appropriate responses accordingly.

REFERENCES

1. R. Kaur and J. Kaur Sandhu, "A Study on Security Attacks in Wireless Sensor Network," 2021 International Conference on Advanced Computing and Innovative Technologies in Engineering (ICACITE), 2021, pp. 850–855. doi: 10.1109/ICACITE51222.2021.9404619
2. Ž. Gavrić and D. Simić, "Overview of DOS Attacks on Wireless Sensor Networks and Experimental Results for Simulation of Interference Attacks," Ingeniería e Investigación, vol. 38, n. 1, pp. 130–138, April 2018.
3. Z. E. Ahmed, R. A. Saeed, and A. Mukherjee, "Challenges and Opportunities in Vehicular Cloud Computing," in Cloud Security: Concepts, Methodologies, Tools, and Applications, 2019, IGI Global, Hershey, PA, 2168–2185. https://doi.org/10.4018/978-1-5225-8176-5.ch106
4. H.-N. Dai, H. Wang, H. Xiao, X. Li and Q. Wang, "On Eavesdropping Attacks in Wireless Networks," 2016 IEEE Intl Conference on Computational Science and Engineering (CSE) and IEEE Intl Conference on Embedded and Ubiquitous Computing (EUC) and 15th Intl Symposium on Distributed Computing and Applications for Business Engineering (DCABES), 2016, pp. 138–141. doi: 10.1109/CSE-EUC-DCABES.2016.173
5. E. Nasr and I. Shahrour, "Evaluating wireless network vulnerabilities and attack paths in smart grid comprehensive analysis and implementation," *2017 Sensors Networks Smart and Emerging Technologies (SENSET)*, 2017, pp. 1–4. doi: 10.1109/SENSET.2017.8125032
6. C. P. Kohlios and T. Hayajneh, "A Comprehensive Attack Flow Model and Security Analysis for Wi-Fi and WPA3," Electronics, vol. 7, pp. 284, 2018. doi:10.3390/electronics7110284
7. S. Kukliński, M. Czerwiński, J. Bieniasz, K. H. Kaminska, D. Paczesny and K. Szczypiorski, "Evaluation of Privacy and Security of GSM Using Low-Cost Methods," *World Conference on Computing and Communication Technologies (WCCCT)*, 2020, pp. 96–105. doi: 10.1109/WCCCT49810.2020.9170000
8. I. C. Eian, K. Y. Lim, M. X. L. Yeap, H. Q. Yeo, and F. Zahra, "Wireless Networks: Active and Passive Attack Vulnerabilities and Privacy Challenges," Oct 2020. doi:10.20944/preprints202010.0018.v1
9. R. Mokhtar and R. A. Saeed, "Conservation of Mobile Data and Usability Constraints," In Z. Junaid and M. Athar (Ed.), Cyber Security Standards, Practices and Industrial Applications: Systems and Methodologies, 2011, IGI Global, Hershey, PA, pp. 40–55.
10. M. K. Hasan, T. M. Ghazal, R. A. Saeed, B. Pandey, H. Gohel, A. A. Eshmawi, S. Abdel-Khalek, and H. M. Alkhassawneh, "A Review on Security Threats, Vulnerabilities, and Counter Measures of 5G Enabled Internet-of-Medical-Things," IET Communications, vol. 16, pp. 1–21, 2021. doi: https://doi.org/10.1049/cmu2.12301

11. E. S. Ali Ahmed and R. A. Saeed, "A Survey of Big Data Cloud Computing Security," International Journal of Computer Science and Software Engineering (IJCSSE), vol. 3, no. 1, pp. 78–85, December 2014.

12. M. M. Saeed, M. K. Hasan, A. J. Obaid, R. A. Saeed, R. A. Mokhtar, E. S. Ali, M. Akhtaruzzaman, S. Amanlou, A. K. M. Zakir Hossain, "A Comprehensive Review on the Users' Identity Privacy for 5G Networks," IET Communications, pp. 1–16, 2022. doi.org/10.1049/cmu2.12327

13. M. M. Saeed, R. A. Saeed, and E. Saeid, "Survey of Privacy of User Identity in 5G: Challenges and Proposed Solutions," Saba Journal of Information Technology and Networking (SJITN), vol. 7, no. 1, 2019.

14. A. Khraisat, I. Gondal, P. Vamplew, et al., "Survey of Intrusion Detection Systems: Techniques, Datasets and Challenges," Cybersecurity, vol. 2, pp. 20, 2019. https://doi.org/10.1186/s42400-019-0038-7

15. B. Subba, S. Biswas, and S. Karmakar, "Intrusion Detection Systems using Linear Discriminant Analysis and Logistic Regression," 2015, IEEE.

16. G. Creech and J. Hu, "Generation of a new IDS test dataset: 'Time to retire the KDD collection,'" IEEE Conference on Wireless Communications and Networking (WCNC), 2013; p. 4487–4492.

17. K. S. Elekar, National Informatics Centre, "Combination of Data Mining Techniques for Intrusion Detection System," IEEE International Conference on Computer, Communication and Control (IC4), pp. 1–5, 2015.

18. A. Chauhan, G. Mishra, and G. Kumar, "Survey on Data Mining Techniques in Intrusion Detection," International Journal of Scientific & Engineering Research, vol. 2, no. 7, July-2011.

19. S.M. Othman, F.M. Ba-Alwi, N.T. Alsohybe, et al., "Intrusion Detection Model using Machine Learning Algorithm on Big Data Environment," Journal of Big Data, vol. 5, pp. 34, 2018. https://doi.org/10.1186/s40537-018-0145-4

20. S. Mukkamala, A. H. Sung, and A. Abraham, "Intrusion Detection Using an Ensemble of Intelligent Paradigms," Journal of Network and Computer Applications, vol. 28, no. 2, pp. 167–182, April 2005. https://doi.org/10.1016/j.jnca.2004.01.003

21. N. Ashraf, W. Ahmad, and R. Ashraf, "A Comparative Study of Data Mining Algorithms for High Detection Rate in Intrusion Detection System (2018)," Annals of Emerging Technologies in Computing (AETiC), vol. 2, no. 1, pp. 49–57, 1st January 2018.

22. X. Sun, D. Zhang, H. Qin, and J. Tang, "Bridging the Last-Mile Gap in Network Security via Generating Intrusion-Specific Detection Patterns through Machine Learning," Security and Communication Networks, vol. 2022, Article ID 3990386, pp. 20, 2022. https://doi.org/10.1155/2022/3990386

23. N. T. Pham, E. Foo, S. Suriadi, H. Jeffrey, and H. F. M. Lahza, "Improving Performance of Intrusion Detection System Using Ensemble Methods and Feature Selection," Association for Computing Machinery, Brisbane, Australia, 2018.

24. U. S. K. Perera Miriya Thanthrige, J. Samarabandu, and X. Wang, "Machine Learning Techniques for Intrusion Detection on Public Dataset," IEEE Canadian Conference on Electrical and Computer Engineering (CCECE), 2016.

25. M. Zamani and M. Movahedi, "Machine Learning Techniques for Intrusion Detection," arXiv:1312.2177v2 [cs.CR] May 9, 2015.

26. P. Gupta and S. K. Shinde, "Genetic algorithm technique used to detect intrusion detection," in Advances in Computing and Information Technology, 2011, Springer, pp. 122–131.

27. S. Mukkamala, G. Janoski, and A. Sung, "Intrusion detection using neural networks and support vector machines," Proceedings of the 2002 International Joint Conference on Neural Network (IJCNN), vol. 2, pp. 1702–1707, 2002.

28. P. Mishra, V. Varadharajan, U. Tupakula, and E. S. Pilli, "A Detailed Investigation and Analysis of using Machine Learning Techniques for Intrusion Detection," *IEEE Communications Surveys & Tutorials*, 2018.
29. J. Kim, P. J. Bentley, U. Aickelin, J. Greensmith, G. Tedesco, and J. Twycross, "Immune System Approaches to Intrusion Detection a Review," Natural Computing, vol. 6, no. 4, pp. 413–466, December 2007.
30. S. Thaseen and Ch. A. Kumar, "An Analysis of Supervised Tree Based Classifiers for Intrusion Detection System," Proceedings of the 2013 International Conference on Pattern Recognition, Informatics and Mobile Engineering, Feb 2013.
31. A. Alazab, M. Hobbs, J. Abawajy, and M. Alazab, "Using Feature Selection for Intrusion Detection System," *International Symposium on Communications and Information Technologies*, 2012.
32. K. Rai, M. Syamala Devi, and A. Guleria, "Decision Tree Based Algorithm for Intrusion Detection," International Journal of Advanced Networking and Applications, vol. 07, no. 04, pp. 2828–2834, 2016.
33. S. Sahu and B. M. Mehtre, "Network Intrusion Detection System Using J48 Decision Tree," 2015, IEEE.
34. W. L. Al-Yaseen, Z. A. Othman, and M. Z. Ahmad Nazri, "Multi-Level Hybrid Support Vector Machine and Extreme Learning Machine Based on ModiÞed K-means for Intrusion Detection System," Expert Systems With Applications, vol. 67, pp. 296–303, 2016.
35. R. Vijayanand, D. Devaraj, and B. Kannapiran, "Intrusion Detection System for Wireless Mesh Network Using Multiple Support Vector Machine Classifers with Genetic-Algorithm-Based Feature Selection", Computers & Security, 2018. doi: 10.1016/j.cose.2018.04.010
36. S. Peng, Q. Hu, Y. Chen, and J. Dang, "Improved Support Vector Machine Algorithm for Heterogeneous Data," Pattern Recognition, vol. 48, no. 6, pp. 2072–2083, 2014. doi:10.1016/j.patcog.2014.12.015
37. E. A. Shams and A. Rizaner, "A Novel Support Vector Machine Based Intrusion Detection System for Mobile ad hoc Networks," Wireless Networks, vol. 24, pp. 1821–1829, 2017.
38. B. S. Bhati and C. S. Rai, "Analysis of Support Vector Machine-based Intrusion Detection Techniques," Arabian Journal for Science and Engineering, vol. 24, pp. 2371–2383, 2019
39. N. N. P. Mkuzangwe and F. V. Nelwamondo, "A Fuzzy Logic Based Network Intrusion Detection System for Predicting the TCP SYN Flooding Attack," 2017, Springer International Publishing.
40. W. Haider, J. Hua, J. Slay, B.P. Turnbull, and Y. Xie, "Generating Realistic Intrusion Detection System Dataset Based on Fuzzy Qualitative Modeling," Journal of Network and Computer Applications, vol. 87, pp. 185–192, 2017.
41. E. S. Ali, M. K. Hasan, R. Hassan, R. A. Saeed, M. B. Hassan, S. Islam, N. S. Nafi and S. Bevinakoppa, "Machine Learning Technologies for Secure Vehicular Communication in Internet of Vehicles: Recent Advances and Applications," Wiley-Hindawi, Journal of Security and Communication Networks (SCN), vol. 2021, 2021. https://doi.org/10.1155/2021/8868355
42. R. A. Saeed, M. M. Saeed, R. A. Mokhtar, H. Alhumyani, and S. Abdel-Khalek, "Pseudonym Mutable Based Privacy for 5G User Identity," Journal of Computer Systems Science and Engineering, vol. 29, no. 1, pp. 1–14, 2021.
43. F. Gumus, C. Okan Sakar, Z. Erdem, and O. Kursun, "Online Naive Bayes Classification for Network Intrusion Detection," IEEE/ACM International Conference on Advances in Social Networks Analysis and Mining, pp. 670–674, 2014.
44. Y. Yu and H. Wu, "Anomaly Intrusion Detection Based Upon Data Mining Techniques and Fuzzy Logic," IEEE International Conference on Systems, Man, and Cybernetics October 14–17, 2012, Seoul, Korea.

45. P. Ghosh and R. Mitra, "Proposed GA-BFSS and Logistic Regression based Intrusion Detection System," 2015, IEEE.

46. G. P. Gupta and M. Kulariya, "A Framework for Fast and Efficient Cyber Security Network Intrusion Detection using Apache Spark," 6th International Conference on Advances in Computing & Communications, Sept. 6–8 2016, Cochin, India.

47. S. A. Mahboub, E. Sayed Ali Ahmed, and R. A. Saeed, "Smart IDS and IPS for Cyber-Physical Systems," in A. Luhach, & A. Elçi (Ed.), Artificial Intelligence Paradigms for Smart Cyber-Physical Systems, 2021, IGI Global, pp. 109–136. https://doi.org/10.4018/978-1-7998-5101-1.ch006

48. M. C. Belavagi and B. Muniyal, "Performance Evaluation of Supervised Machine Learning Algorithms for Intrusion Detection," Procedia Computer Science, vol. 89, pp. 117–123, 2016.

49. E. Besharati, M. Naderan, and E. Namjoo, "LR-HIDS: Logistic regression host based intrusion detection system for cloud environments," Springer-Verlag GmbH Germany, Springer Nature, 2018.

50. M. N. Aziz and T. Ahmad, "Cluster Analysis-Based Approach Features Selection on Machine Learning for Detecting Intrusion," International Journal of Intelligent Engineering and Systems, vol. 12, no. 4, 2019.

51. L. Tan, C. Li, J. Xia, and J. Cao, "Application of Self-Organizing Feature Map Neural Network Based on K-means Clustering in Network Intrusion Detection," Computers, Materials & Continua, vol. 61, no.1, pp. 275–288, 2019.

52. S. Duque and M. N. bin Omar, "Using Data Mining Algorithms for Developing a Model for Intrusion Detection System (IDS)," Complex Adaptive Systems, Publication 5, Conference Organized by Missouri University of Science and Technology, 2015, San Jose, CA.

53. O.Y. Al-Jarrah, Y. Al-Hammdi, P.D. Yoo, S. Muhaidat, and M. Al-Qutayri, "Semisupervised Multi-Layered Clustering Model for intrusion detection," Digital Communications and Networks, 2017. doi: 10.1016/j.dcan.2017.09.009

54. F. Chen, Z. Ye, C. Wang, L. Yan, and R. Wang, "A Feature Selection Approach for Network Intrusion Detection Based on Tree-Seed Algorithm and K-Nearest Neighbor," The 4th IEEE International Symposium on Wireless Systems within the International Conferences on Intelligent Data Acquisition and Advanced Computing Systems, Lviv, Ukraine, Sept. 20–21, 2018, 68–72.

55. A. R. Syarif and W. Gata, "Intrusion Detection System using Hybrid binary PSO and k-nearest Neighbourhood Algorithm," International Conference on Information & Communication Technology and System (ICTS), pp. 181–186, 2017.

56. K. S. Desale and R. Ade, "Genetic Algorithm based Feature Selection Approach for Effective Intrusion Detection System," 2015 International Conference on Computer Communication and Informatics, Jan. 8–10, 2015, Coimbatore, India.

57. P. A. A. Resende and A. C. Drummond, "Adaptive anomaly-based intrusion detection system using genetic algorithm and profiling," *Security and Privacy*, vol. 1, no. 4 2018. doi: 10.1002/spy2.36

58. M. Manimekalai and G. Anupriya, "A Novel Intrusion Detection System using Data Mining Techniques," Journal of Emerging Technologies and Innovative Research, vol. 6, no. 6, June 2019.

59. M. A. Alsheikh, S. Lin, D. Niyato, and H.-P. Tan, "Machine Learning in Wireless Sensor Networks: Algorithms, Strategies, and Applications," IEEE Commun. Surv. Tutorials, vol. 16, no. 4, pp. 1996–2018, 2014.

60. F. Gharibian and A. A. Ghorbani, "Comparative Study of Supervised Machine Learning Techniques for Intrusion Detection," in Fifth Annual Conference on Communication Networks and Services Research (CNSR '07), pp. 350–358, 2007.

61. E. Hodo, X. Bellekens, A. Hamilton, P.-L. Dubouilh, E. Iorkyase, C. Tachtatzis, and R. Atkinson, "Threat analysis of IoT networks Using Artificial Neural Network Intrusion Detection System," 2016, IEEE.1

62. F. Amato, N. Mazzocca, E. Vivenzio, and F. Moscato, "Multilayer Perceptron: an Intelligent Model for Classification and Intrusion Detection," 31st International Conference on Advanced Information Networking and Applications Workshops, pp. 686–691, 2017.

63. W. A. H. M. Ghanem and A. Jantan, "A New Approach For Intrusion Detection System Based on Training Multilayer Perceptron by Using Enhanced Bat Algorithm" Springer-Verlag London Ltd., part of Springer Nature, 2019.

64. R. A. Mokhtar, S. Khatun, A. Borhanuddin, A. Ramli, and R. A. Saeed, "Authentication and User Presence Monitoring Technique for Mobile Computers Using JSR82," Proceeding of Brunei International Conference on Engineering and Technology (BICET 2005), July 2005, vol. (2), pp. 133–136, Bandar Seri Begawan, Brunei.

65. M. Saeed, R. A. Saeed, and E. Saeid, "Identity Division Multiplexing Based Location Preserve in 5G," IEEE International Conference of Technology, Science and Administration (ICTSA), 2021. doi: 10.1109/ICTSA52017.2021.9406554

66. M. M. Saeed, R. A. Saeed, and E. Saeid, "Preserving Privacy of Paging Procedure in 5thG Using Identity-Division Multiplexing", First International Conference of Intelligent Computing and Engineering, Yemen, 2019, pp. 1–6.

67. X. Qu, L. Yang, K. Guo, L. Ma, M. Sun, M. Ke, M. Li, "A Survey on the Development of Self-Organizing Maps for Unsupervised Intrusion Detection," Springer Science+ Business Media, part of Springer Nature, 2019.

68. B. J. Bejoy and S. Janakiraman, "Artificial Immune System Based Intrusion Detection System-A Comprehensive Review," International Journal of Computer Engineering & Technology (IJCET), vol. 8, no. 1, Jan.–Feb. 2017.

69. D. Hooks, X. Yuan, K. Roy, A. Esterline, and J. Hernandez, "Applying Artificial Immune System for Intrusion Detection," IEEE Fourth International Conference on Big Data Computing Service and Applications, 2018.

70. C. Yin, L. Ma, and L. Feng, "Towards Accurate Intrusion Detection Based on Improved Clonal Selection Algorithm," *Springer Science+Business Media New York*, 2015.

71. O. Elezaj, S. Y. Yayilgan, M. Abomhara, P. Yeng, and J. Ahmed, "Data-driven Intrusion Detection System for Small and Medium Enterprises," 2019, IEEE.

72. S. Agrawal, S. Sarkar, O. Aouedi, G. Yenduri, K. Piamrat, S. Bhattacharya, P. K. Reddy Maddikunta, and T. R. Gadekallu, "Federated Learning for Intrusion Detection System: Concepts, Challenges and Future Directions," arXiv:2106.09527v1 [cs.CR] 16 Jun 2021.

73. S. Anwar, J. M. Zain, M. F. Zolkipli, Z. Inayat, S. Khan, B. Anthony, and V. Chang, "From Intrusion Detection to an Intrusion Response System: Fundamentals, Requirements, and Future Directions," Algorithms, vol. 10, pp. 39, 2017. doi:10.3390/a10020039

74. C. Briggs, Z. Fan, and P. Andras, "Federated Learning With Hierarchical Clustering of Local Updates to Improve Training on non-IID Data," International Joint Conference on Neural Networks (IJCNN), pp. 1–9. IEEE, 2020.

75. Q. Qin, K. Poularakis, K. K. Leung, and L. Tassiulas, "Line-speed and Scalable Intrusion Detection at the Network Edge Via Federated Learning," IFIP Networking Conference (Networking), pp. 352–360. IEEE, 2020.

76. Y. Fan, Y. Li, M. Zhan, H. Cui, and Y. Zhang, "IoTDefender: A Federated Transfer Learning Intrusion Detection Framework for 5G IoT," IEEE 14th International Conference on Big Data Science and Engineering (BigDataSE), pp. 88–95. IEEE, 2020.

77. J. Ren, H. Wang, T. Hou, S. Zheng, and C. Tang." Federated Learning-Based Computation Offloading Optimization in Edge Computing-Supported Internet of Things," IEEE Access, vol. 7, pp. 69194–69201, 2019.

78. S. Latif, Z. Zou, Z. Idrees, and J. Ahmad, "A Novel Attack Detection Scheme for the Industrial Internet of Things Using a Lightweight Random Neural Network," IEEE Access, vol. 8, pp. 89337–89350, 2020.

79. A. R. Short, H. C. Leligou, M. Papoutsidakis, and E. Theocharis, "Using Blockchain Technologies to Improve Security in Federated Learning Systems," IEEE 44th Annual Computers, Software, and Applications Conference, pp. 1183–1188, 2020. doi: 10.1109/COMPSAC48688.2020.00-96

80. H. Zhu and Y. Jin, "Multi-objective Evolutionary Federated Learning," IEEE Transactions on Neural Networks and Learning Systems, vol. 31, no. 4, pp. 1310–1322, 2019.

81. T. U. Chengsheng, L. I. U. Huacheng, and X. U. Bing, "AdaBoost Typical Algorithm and Its Application Research," MATEC Web of Conferences, vol. 139, 00222, 2017. doi: 10.1051/matecconf/201713900222

82. P. Albertos, A. Sala, and M. Olivares, "Fuzzy Logic Controllers. Methodology, Advantages and Drawbacks," Conference Paper, September 2000. doi: 10.13140/RG.2.1.2512.6164

83. G. Karatas, O. Demir, and O. Koray Sahingoz, "Deep Learning in Intrusion Detection Systems," International Congress on Big Data, Deep Learning and Fighting Cyber Terrorism, 2018, pp. 113–116. doi: 10.1109/IBIGDELFT.2018.8625278

84. R. Vinayakumar, M. Alazab, K. P. Soman, P. Poornachandran, A. Al-Nemrat, and S. Venkatraman, "Deep Learning Approach for Intelligent Intrusion Detection System," in IEEE Access, vol. 7, pp. 41525–41550, 2019. doi: 10.1109/ACCESS.2019.2895334

85. K.-A. Tait, J. S. Khan, F. Alqahtani, A. A. Shah, F. A. Khan, and M. U. Rehman, "Intrusion Detection using Machine Learning Techniques: An Experimental Comparison," arXiv:2105.13435v1 [cs.CR] 27 May 2021.

86. H. Choi, M. Kim, G. Lee, W. Kim, "Unsupervised Learning Approach for Network Intrusion Detection System Using Autoencoders," Springer Science+Business Media, part of Springer Nature, 2019.

87. G. Caminero, M. Lopez-Martin, and B. Carro, "Adversarial Environment Reinforcement Learning Algorithm for Intrusion Detection," Computer Networks, vol. 159, pp. 96–109, 2019. doi: https://doi.org/10.1016/j.comnet.2019.05.013

88. M. Lopez-Martin, B. Carro, and A. Sanchez-Esguevillas "Application of Deep Reinforcement Learning to Intrusion Detection for Supervised Problems," Expert Systems with Applications, vol. 141, p. 112963, 2020. doi: https://doi.org/10.1016/j.eswa.2019.112963

89. M. Sarnovsky and J. Paralic, "Hierarchical Intrusion Detection Using Machine Learning and Knowledge Model," Symmetry, vol. 12, pp. 203, 2020. doi:10.3390/sym12020203

Part III

Advanced Topics in Data-Driven Intelligence for Wireless Networks

8 Policy-based Data Analytic for Software Defined Wireless Sensor Networks

Rashid Amin

University of Engineering of Technology, Taxila,
Pakistan; University of Chakwal, Chakwal, Pakistan

Mudassar Hussain

Department of Computer Science,
University of Sialkot, Pakistan

Saima Bibi

Department of Computer Science, University of
Engineering and Technology, Taxila, Pakistan

Ayesha Sabir

Department of Computer Science, University of
Engineering and Technology, Taxila, Pakistan

CONTENTS

DOI: 10.1201/9781003216971-11

189

8.1   INTRODUCTION

Software defined networking (SDN) [1–4] is an emerging type of network that separates network control and management from the data plane to reduce complexity and increase network management. This separation of the control and data planes leaves network switches as simple forwarding devices. However, the network control is shifted to the centralized logical entity called the controller, which acts as the network brain and maintains a global view of the network and programming abstractions. It offers programmatic control of the entire network and provides network operators real-time control of underlying devices. The management plane specifies network applications, such as network policies, network monitoring, load balancing, etc. These are implemented by the network administrator based on the application environment and user's requirements. In this way, network management turns out to be simple, reducing network rigidness.

8.2   APPLICATION PROGRAMMING INTERFACES

The three SDN planes interact with each other using application programming interfaces (APIs) [5]. APIs are very important in SDN, as they provide communication between the data, control, and management planes. The well-known APIs are Southbound APIs (SBI), Northbound APIs (NBI), and in the case of distributed controllers, East/Westbound APIs. These APIs are an architectural component of SDN and are used to configure forwarding devices or network applications. The SDN layered architecture, including APIs, is shown in Figure 8.1.

8.2.1   Southbound APIs

The SBI is an SDN enabler that provides a communication protocol between the control plane and the data plane. It is used to push configuration information and install flow rules at the data plane. It also provides an abstraction of the network device's functionality to the control plane. SBI is critical with respect to its availability as well as secure communication. In the absence of any one of these parameters, it may result in the malfunctioning of forwarding devices. It faces heterogeneity challenges, vendor-specific forwarding devices, and programming languages. OpenFlow Protocol (OFP) is commonly used as an SBI that provides a secure channel between the control and data planes to install flow rules. However, it includes multiple OpenFlow dependent and independent SBIs. It also comprises OpenFlow-based SBIs in emerging technology. Sensor OpenFlow (SOF) [6] is based on standard OpenFlow, but modified to the requirements of low-capacity sensor nodes. It addresses challenges like: flow creation, secured channels between the control plane and data plane, control traffic overhead,

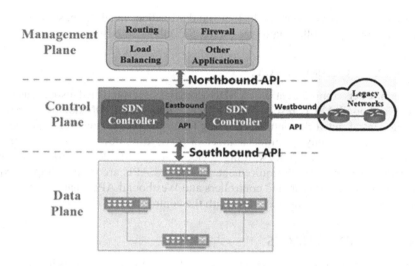

FIGURE 8.1   SDN system architecture.

and in-network processing, etc. Installing the flows on sensor network devices suggests the redefining of flow tables due to the special addressing schemes of wireless sensor networks (WSNs). Software defined wireless networks (SDWNs) [7] aim to decrease energy consumption in WSNs with the help of duty cycles and in-network data aggregation. Duty energy minimizes consumption by turning radio frequencies off in case of an idle period, and in-network data aggregation is also helpful. Its protocol architecture utilizes both generic and sink nodes. In addition, it supports elastic flow rules due to its nature, unlike traditional OpenFlow. SDN for wireless sensors (SDN-WISE) [8] implemented in OMNet++ aims to reduce sensor node communication with the controller and make sensor nodes programmable. In the control plane, it utilizes a topology manager to collect local information from the nodes and forwards it to the controller. In the data plane, in-networking packet processing is responsible for data aggregation and other in-network processing to reduce the overhead. The adaptation layer is responsible for communication between the control and data planes. SOF and SDWN provide theoretical details using a centralized controller, whereas SDN-WISE provides practical implementations using the ONOS controller [9], which is distributed. Nodes in sensor networks are susceptible to movement, which can cause path variation during packet transmission. It is very important to manage and monitor the movement of different nodes. One of the major challenges of SDN is to handle the effects of nodes entering or leaving the network. Another challenge is to build paths using different metrics (i.e., node energy and capability). The interface in such cases should be able to gather required information for the controller.

8.2.2 Northbound APIs

Northbound API (NBI) is important concerning SDN adoption's ability to support a variety of SDN applications. Current controllers and programming languages offer a wide range of NBIs due to the lack of standardization. In addition, some programmers and many controller platforms use REST API as an NBI.

8.2.3 EAST/WESTBOUND APIs

Inter-controller communication of SDN domains is established using Eastbound API. Westbound API is responsible for legacy-domain-to-SDN-domain communication. Central network control is the key feature of SDN. However, a single controller can handle only a limited number of forwarding devices. To accommodate the exponential increase in forwarding devices and large-scale networks, distributed controllers have become a requirement. Eastbound APIs are used to import/export information among distributed controllers and Westbound APIs enable communication between legacy network devices with the controllers.

8.3 SDN ADVANTAGES

SDN has many advantages compared to the traditional networking paradigm [10–16]. A few of them are listed next:

- SDN increases network control with higher speed and flexibility. Developers program the open standard-based controller to control the flow of traffic over a network. Network managers have many options when selecting the pieces of equipment because they use a single protocol to communicate with many central control hardware devices.
- SDN offers customizable network infrastructure. Administrators oversee creating network services. A software-defined network may assign resources to update the network in real-time. It allows network administrators to improve data flow throughout the network. It improves applications that require a longer response time.
- A SDN can provide a more complete view of security risks. Additionally, it provides visibility across the whole network and many other advantages over traditional networking. The number of Internet gadgets is increasing; an SDN allows operators to create separate zones for devices. It necessitates a higher level of security.
- SDN allows network administrators to apply Access Control List (ACL) policies more granularly in a highly abstracted automated fashion.
- It provides a better user experience by centralizing network control and making state information available to higher-level applications as it can adopt dynamic user needs easily.
- It provides centralized security control, which improves network visibility through security management. In addition, it offers robust control over network infrastructure to develop efficient and effective security mechanisms to detect and prevent security attacks.
- SDN offers mobility support for the Internet of Vehicles (IoV) by enabling intelligent remote clouds for the computation of tasks offloaded by the fast vehicle toward the roadside unit (RSU). The controller supports the RSU

for implementing the communication path among the RSU and fog node with adequate resources in a predictive fashion. Consequently, the fog node capabilities log at the controller tends to the optimal computation of IoV jobs by fog nodes.

- Future networks such as 6G require SDN-enabled softwarization and management for remote and machine-learning application decision-aware reconfigurations in network resources.
- Softwarized policy implementation architecture can enhance the security among autonomous systems that have the least human interactions and, consequently, it can mitigate the security risk of inter-domain communication.

8.4   SDN HISTORY AND EVOLUTION

Traditional networks are complex and hard to manage because all the functionalities of data; the control and management planes are vertically and tightly coupled in forwarding devices [17]. In traditional networks, the control plane with the help of routing protocols forwards data packets as per network policies. Due to this vertical integration and tightly coupled nature of forwarding devices, network management becomes difficult and challenging. Moreover, the network applications and services of the current information age have become more complex and demanding, requiring the Internet to evolve and address these new challenges. To resolve such issues, the idea of "programmable networks" had been proposed to facilitate network evolution. In this regard, two concepts, active networking [18] and programmable networks [19] with respect to network programmability, were explored. Active networking involves network intelligence instead of typical packet processing where network nodes can perform customized operations on packets. The programmable networks permit the controlling of network node behavior and flow control through software. Later, the 4D project [20–22] proposed a clean-slate design that is based on four planes: decision, dissemination, discovery, and data. It emphasizes the separation of routing decision logic and protocols governing the interaction between forwarding devices. The decision plane has a network-wide view of network topology and installs configuration commands for communication at the data plane. The dissemination and discovery planes provide efficient communication between the decision and data planes. Ethane [23] proposed a new network architecture for enterprise that allows network managers to configure and control the whole network using a centralized controller. This research proposed the clear foundation for separation between data and the control plane, which resulted in the introduction of SDN. SDN provides a real-world implementation of a suite of networking software that allows a network to be centrally controlled. It is not the first and only solution that accepts separation and programmability. However, it has wide acceptance in academia and industry due to the rapid innovation in the control and data planes. A group of network operators, service providers, and vendors have created the Open Network Foundation (ONF) [24], which is an industrial-driven organization to promote SDN and standardize the OFP [25]. On the academic side, the OpenFlow Network Research Center [26] has been created with a focus on SDN research.

8.5 WIRELESS SENSOR NETWORKS

WSNs are a collection of devices that can wirelessly relay data obtained from a monitored field. The data is sent through several nodes and connected to other networks via a gateway, such as wireless Ethernet. It is made up of spatially distributed sensors, and one or more sink nodes (also called base stations). These networks are used to display physical or environmental parameters, such as sound, pressure, and temperature and convey data to a central point via the network [27]. WSNs use multiple network topologies based on implementation scenarios. Usually, star and tree mesh topologies are used to build a WSN [28]. The environment determines the category of networks and can be deployed underwater, underground, on land, etc. [29]. In WSN, the sensor nodes are used for constant sensing, event ID, event detection, and local control of actuators. There are a variety of WSN applications in environment tracking, industrial process monitoring, automated building climate control, ecosystem and habitat monitoring, civil structure health monitoring, etc. [30]. These networks use different security algorithms as per underlying wireless technologies to provide a reliable network.

8.5.1 SOFTWARE DEFINED WIRELESS SENSOR NETWORK

A SDWSN aimed to get benefits from SDN for ease of management and central control. SDWSNs help to conserve less energy and sensor node operation is made simpler with the help of central controller. All the Northbound, Eastbound, Westbound, and Southbound APIs of the SDN help to design WSNs for effective central control and ease of management. Many recommendations have been made to improve control-plane function management. System conformation and topology management are at the heart of SDN-based WSN management. The aim of the data plane interface with controller nodes and peripherals of network configuration and topology management is to ensure the network runs efficiently. The application layer and control and data planes are three layers that provide a management system for the Internet of Things (IoT). The network configuration and centralized device settings are based on the architecture controller design. SOF is a core component of communication between two planes such that data plane is flow-based and the control plane is centralized in the network. WSNs employ the same channel for node-to-node communication and data processing due to their limited resource availability. SDN enables the separation of the control plane from the data plane. The control and data packets for WSNs will still employ the current dynamic and resource constrained WSN architecture. The method used for managing the network topology should be able to accommodate node mobility, which leads to changes in topology and handling faulty wireless links. These should be done in conjunction with Quality of Service (QoS) management, which makes sure the network is stable. When managing node localization, localization plans have been proposed to recover the correctness of finding nodes. In SDWSNs, energy management use in the network creates a more efficient system. It's also worth noting that sensor node design focuses on reducing energy consumption by including low-power devices into the automated architecture though maintaining the essential QoS [31].

8.6   SDWSN ARCHITECTURE

There are three logical planes of the SDN that require extracting and reorganizing distinct capabilities of the WSN, such as application, service, network, data, and control. SDN principles include:

8.6.1   Forwarding Devices

The component of a network that transports the data plane. This facilitates data transport to and from clients and manages many communications across many rules and remote peers. This software layer manages the data plane in cloud computing on storage such as drives in addition to native data services such as snapshots, clones, and replication. The data plane includes firewall security features with next generation firewall is responsible for processing flow. The control and management planes serve the data plane, which bears the network traffic transmissions. Forwarding devices are sensor nodes that comprised of hardware and software components. hardware makes up the power unit, sensor, and radio devices. The hardware components are situated in the Physical (PHY) layer, which works in conjunction with the Media Access Control (MAC) layer.

- **MAC and PHY**
  A device's MAC address is its physical location. The PHY layer defines the physical and electrical features of the network and is in charge of the circuitry that modulates and demodulates RF signals. The MAC layer is in charge of sending and receiving radio frequency frames. The MAC/PHY design does not include any beacons or master/slave needs. The PHY, MAC, and network system are three layers of nodes. A generic node is made up of a sink node and a controller.
- **Data Processing**
  Energy is the best significant resource in WSNs. In-network data processing is a typical plan. An intermediate proxy node is selected to house a possibly complex data transformation function to edit sensor data streams enroute to the sink node. The three basic types of data processing are manual, mechanics, and electronic. Data processing is handled manually. The logical activities are data gathering, filtering, sorting, and calculating by humans without using any other technological equipment. It results in a high number of errors, significant labor expenses, a lot of time, and requires a low-cost method.

8.6.2   Controller

A controller is a device that installs rules at network nodes for data flow between two entities, as shown in Figure 8.2. A network interface card is a circuit card that allows a computer to connect to the Internet. Any apparatus that reduces the number of pollutants is a control device. The ratio of the pollutant emitted by a control device to the pollution delivered to the control device is known as control device efficiency. The sink node uses a wireless interface to link with new sensor nodes. The embedded

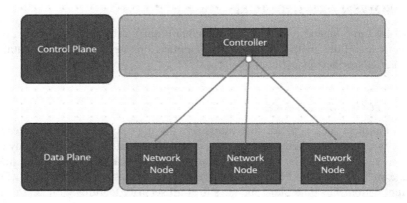

FIGURE 8.2   Control and data planes.

system comprised of an adaptability module, virtualizer, and a controller. The adaptability module handles message formatting. The virtualizer guarantees that all collected data is compiled into a virtualization for transmission to the controller. A main disadvantage of this tool is that the serial connection between the controller and the sink nodes [32].

8.7   SDWSN CHALLENGES

SDWSNs face several challenges, such as network/energy management, security, routing, etc. The SDN approach to WSNs tries to resolve these challenges and foster efficiency and sustainability in WSNs.

- **Energy Management in SDWSN**
  SDWSNs suggested a multi-tasking sensor that prioritizes energy efficiency in routing techniques. The Particle Swarm Optimization (PSO) method was used to choose control nodes in each cluster, dynamically distributing distinct sensing duties to cluster members. Each control node is connected to the controller via a direct link. The TDMA scheduler directs the data to the control node before sending it to the controller. The network lifetime of SDWSN was developed through the optimization algorithm by selecting the optimum number of control nodes that allocate shared tasks in clusters and packets sent to the sink node. However, routing would become more complicated with billions of interconnected devices, and security would remain a big worry. The least energy sensor activation problem is recast as a mixed-integer quadratic programming problem by accounting for sensor activation and task mapping. SDWSN is a good approach to improve the efficiency and sustainability of WSN and to foster interoperability with other networks. [33].
- **Network Management in SDWSN**
  Enterprise management necessitates three distinct components: metrics, monitoring, and systems management. A proper communication network,

which is difficult in the old scheme, is made simple by using the SDN. The end-to-end QoS parameters are implemented by using enumerated finetuned traffic measurements. The network management difficulties are addressed to adapt to changing policies by the WSN. All three of the categories cause packets to be fed to the target node, resulting in a decrease in network performance due to task denial. Distributed denial-of-service (DDoS) attacks target any application in any network, resulting in ongoing commercial losses [34].

- **Routing Management in SDWSN**
  Routing is the process of directing traffic within a network and between and across several networks. Routing takes the public switched telephone network (PSTN) and computer networks, including circuit-switched networks, within a variety of networks. Routing is hub for all IP connectivity. Routing forms basic connections to unique and individual devices into a hierarchical structure that implements an addressing scheme. Routing is a system used by networks to determine the best path for data to take. Network routing considers the different paths that data can travel across a network and selects the best one. Routing is critical in networks because it allows data to be transmitted soon as feasible. Routing can be classified into three types: static, dynamic, and default routing. The routing protocol of mobile node networks through a file-sharing application employs end-to-end communication over critical messages shared by the mobile nodes. These two types of communications each have their own set of issues in terms of enabling reliable and secure data flow between nodes or between nodes and controllers [35].

- **Localization**
  To approximate localization, localized and unlocalized are used to estimate their geometrical position or placement within a communication. This chapter compares and applies three data and control plane localization strategies. The SDN sensor nodes choose the best traffic routing path with the help of the controller node.

- **Security**
  Selective forwarding are malicious attacks that can be mitigated by the Energy-efficient Trust Management and Routing Mechanism (ETMRM). The ETMRM specifically solves two critical concerns with SDWSNs. Better security requires accurate detection and separation of malicious forwarding attack nodes and a highly trustworthy and dependable routing path. The security of topological data collection regularly prevents unwanted repercussions and ensures energy efficiency. A report message that saves electricity, to reduce control overhead and protect control traffic transmission, was developed as a type of secure aggregation method. To address the security difficulties in SDWSNs, the study [36] introduces hierarchical trust management (HTM). Every level of the hierarchical structure examines each node's trustworthiness and detects harmful behavior of the HTM system. Sensor nodes based on their forwarding behavior, offer an energy-efficient, software-defined network-based mobile

code-driven trust mechanism (MCTM) for resolving an issue. The MCTM uses a mobile code and follows pre-determined routes, collecting necessary information to assess their trustworthiness. In addition, a solution presented to mitigate sinkhole assaults is the hop count-based detection strategy [36].

8.8   SOFTWARE DEFINED NETWORK METHODS USED IN WIRELESS SENSOR NETWORK

A multi-thread operating system provides the sensors for WSNs. MANTIS remains an energy-saving operating system. Even though there are still a lot of open study topic networks (WSNs), there are currently a great deal of existing situations where these networks might be used. Tracking, monitoring, surveillance, building automation, military applications, and agriculture are just a few examples of application domains. Various energy consumption in WSNs have been used to overcome the problem. Different approaches can decrease energy usage but message issues and traffic must be managed. Finally, most present Multi-hop Communication with Localization (MCL) based techniques have no significant localization delay in their behavior for time-critical monitoring systems.

- To service sampling and compression procedures, use the distance source coding method.
- Distributed and parallel transaction devices decrease traffic in the stimulating WSN environment.
- The features of effective energy utilization, loop trees, and hop-by-hop routing algorithms are investigated.
- The network's performance may be tested and simulated [37].

8.9   NETWORK POLICIES IN SOFTWARE DEFINED-BASED WIRELESS NETWORKS

Existing techniques have addressed these issues by presenting a variety of alternate strategies for effectively and correctly implementing network policies in hybrid SDN, listed here:

- Release of network policies: Policies can be specified in an SDN architecture using parallel and sequential operators.
- In network policies, there are overlaps and conflicts: Some network policies may declare ACL policy openings, network reachability concerns, packet loss, and ACL policy overlaying and conflict.
- Some are kept at the controller: Various approaches, such as Policy Graph Abstraction (PGA), can be used to store network policies at the controller. When a packet reaches at controller, PGA walks the graph and forms flow rules to implement conflict free ACL policies.
- The ACL policies are a set of switch flow rules used to manage unwanted boundary network traffic.

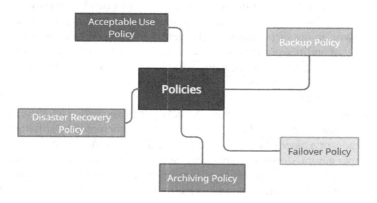

FIGURE 8.3   Types of network policies.

- The network policies connect at perfect points in the network, dipping the general number of functional network policies and undesired transmissions in the network, as the network policies are designed to do.
- The network policies can be installed either pre-emptively or reactively by the controller. The ternary content addressable memory (TCAM) memory problem is produced by the active and passive connecting network policies for all users in a proactive way. The controller responds by connecting network policies exclusively for active users. As a result, the reactive technique makes better use of the switches' limited TCAM memory [38].

8.10   TYPES OF NETWORK POLICIES

A network is an advanced structure that comprises different network devices. The network devices communicate with each other based on network policies. Different types of network policies are shown in Figure 8.3. The network manager will have created these, usually after speaking with everyone. Acceptable use, disaster recovery, back-up, archiving, and failover policies are among them.

- **Acceptable Use Policy**
  People who require access to a network to execute their jobs are frequently asked to sign a contract. They will only use the information for commercial reasons. They must also commit not to use it for illegal, immoral, or time-consuming activities such as downloading copyrighted information, viewing pornography, or spending time on non-business-related social networking sites. This policy ensures that an organization knows who is responsible for the network and its utilization.
- **Disaster Recovery Policy**
  There's a danger that all networks will burn to the ground. This may be anything from a network-connected structure catching fire, to floods, to

deliberate damage committed by a hacker or a disgruntled employee. When a calamity hits, a company's network must be restored as soon as possible, or it will go out of business. A disaster recovery policy is a written document that outlines what should be done in the event of a disaster. Companies that plan and prepare for the worst will know what to do quickly and quietly in the event of a major crisis.

- **Backup Policy**

  All data is in danger at all times. A hacker or a disgruntled employee might destroy data, there could be a fire, data could be corrupted by software or an upgrade, or a piece of equipment like a hard drive could fail. An earthquake or other natural disaster might occur. Someone may, for example, give away a computer with data on it. To ensure safety, data must be backed up regularly. A policy is written down to ensure that it is understood completely.

  - Whose responsibility is it to perform backups?
  - How often and when will backups be made, and how will each backup be created?
  - How will each backup be labeled?
  - Where will backups be kept?
  - How will the backup system be tested?

  Backups are usually done regularly. For example, in the first week of a four-week backup plan, an "incremental backup" may be performed every night on backup tapes named Monday1, Tuesday1, Wednesday1, and Thursday1. In an incremental backup, just the files that have changed since the last backup are resaved. A "full backup" of the whole system's data is produced on a tape called Friday1 every Friday.

  This is then repeated on Monday2, Tuesday2, Wednesday2, Thursday2, and Friday2, for a total of 10 cassettes. The third and fourth weeks are handled in the same way. At the start of the new four-week period, the cases are rewritten, and the process begins all over again.

  The policy outlined here may be sufficient for a small firm. You'd retain backups going back months, if not years for bigger files. The reason for so many backups is that a malfunction with the backup system may take weeks or months to detect. It would be a tremendous problem if you realized you only had one cassette and backed up a computer every day on that one tape. For example, one of the backup tapes was defective or the backup device hadn't been operating correctly for a few weeks.

- **Archiving Policy**

  You don't always want firm files on your system, but you also don't want to delete them. For example, when a class of students leaves school to attend university or begin working, for examples, their information does not have to be retained in the school's administrative system. When a company's tax and accounting records for the year are complete, the business no longer needs them, but they can't be erased in case the IRS wants to look at them. If a plane crashes, the jet's designers may be forced to review their records decades after the plane was built; they don't need the details of these components in their present system, but need them to be available for the future.

The presence of old data on a system delays searches, increases backup times, and depletes hard disc space. As a result, data that is no longer needed is "archived:"

- The information is compressed down to the smallest size feasible
- The information is saved on a storage device, such as online storage or a backup tape
- It is correctly labeled with a date and archival information
- Backups and archiving may appear to be the same thing, but they are not. They are carried out for different reasons and in entirely distinct methods.

- **Failover Policy**
  Companies frequently install backup devices for crucial network equipment in case the original device fails. They are set up to start automatically if the first device fails and show that there is no downtime on the company's network. When a company cannot access its network, this is known as downtime. If a company's network went down for even a few hours, it would be disastrous. It might cost them tens of thousands of pounds/dollars, if not millions. Equipment, such as servers and routers, is an obvious example. This is how it is emulated. Companies usually have a written approach known as a Failover policy, which specifies, among other things, which pieces of equipment will be emulated, how they will be set up, and how long customers may expect downtime [39].

8.11   OPTIMIZATION OF POLICIES

Security and privacy policies are ingrained in both technological and social systems, and technology enables both enterprises and individuals to produce and share information. In IT, the ability to administer such policies is viewed as a vital requirement. Policy authoring is promoted as a vital component of useable privacy and security systems. In a policy management ecosystem with policy templates, multiple tasks with various skillsets are crucial. The Policy Optimizer module makes it simple to move an existing security policy rule base to an App-ID-based rule base, which enhances security by minimizing the attack surface and allowing safety-enabled applications. Policy Optimizer detects port-based rules, allows conversion to application-based rules, or add apps from a port-based rule to an existing application-based rule without endangering application availability. It also detects App-ID-based rules that have been over-provisioned. Policy Optimizer aids in the prioritization of which port-based rules to migrate first, the identification of application-based rules that allow applications that are not in use, and the analysis of rule usage characteristics such as hit count [40]. Converting port-based rules to application-based rules enhances security posture by allowing specific programs and denying all others, removing unwanted and potentially harmful traffic from the network. Converting application-based rules, when combined with confining application traffic to its default ports (changing the service to "application-default"), prohibits evasive programs from running on non-standard ports [41].

8.12 SDN AS A DATA-DRIVEN SOLUTION FOR
WIRELESS NETWORKS

The traditional networks have evolved from academic infrastructure to commercial networks to provide a vital information platform for communications. However, these networks face the problems of complicated management, manual configurations, and security issues. To resolve these issues, SDN introduced a promising networking paradigm that separates network control and management from the forwarding devices. It simplifies network management by providing APIs that help to implement network policies and other applications from the central controller. In this way, it simplifies network management, improves performance, encourages network innovation, and automates configurations. PGA [42] provides automatic and conflict-free policies, for instance, network policies, load balancing, [43] etc. It examines various network policies that are individually stated for any conflict as shown in Figure 8.4.

In different situations, network policies conflict with each other due to various perspectives. The graph composition is very helpful to express conflict-free network policies. As a next step, these policies are forwarded to a graph composer through

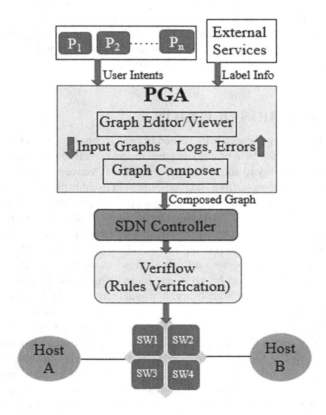

FIGURE 8.4   PGA system architecture presenting phenomena of automatic and conflict free network policies.

a PGA user interface (UI). It resolves conflicts or gives some possible suggestions to the network administration. Finally, it generates error-free/conflict-free graphs. A service function chain-based approach specifies and verifies ACL policies to detect anomalies in ACL policies prior to deployment [44]. To achieve the desired goal, the forwarding policies are formally represented and a set of anomalies are detected against a set of flow rules for the respective policies. In addition, it also provides provision for network administrators to specify their own anomalies. The results state that the proposed approach can verify anomalies of a reasonable-sized network in milliseconds. Moreover, the tools like, FlowChecker [45] and Anteater [46] are useful to troubleshoot, debug, and detect anomalies in communication networks.

The trend of increasing massive IoTs and continuous streaming traffic is steering the demand of increasing computations. Cloud-Fog hybrid systems support delay-intensive applications in a distributed computing manner. SDN supports various network infrastructure and inter-controller communication models (flat, horizontal, vertical, hybrid, or T-model) for distributed network management [47]. The vulnerability and consistency challenge in distributed architecture is more likely to be held in contrast to the central control. Currently to support the applications in 5G networks and beyond, the SDN distributed frameworks need to be more sophisticated. Rahman et al. [48] proposed SmartBlock-SDN for efficient resource management and security assurance in blockchain-enabled IoT networks. The proposed framework addresses the challenges of distributed control security and the energy-efficient cluster head selection in controllers. SmartBlock-SDN is mapped for a layered approach (IoT, edge, cloud) and cloud-enabled blockchain considers coping with the various common network vulnerabilities. Distributed homogeneous controllers and enforced network policies are recorded using immutable blockchains. This stored policy configuration can be accessed using REST API for various operations in line with network security and resources management. BMC-SDN architecture incarcerates the SDN and blockchain in a network, where control is distributed for the fault tolerance and redundant control resources [49]. It employed the blockchain for redundant controllers in various segregated domains. East/West (inter-controllers) communication and network operations are recorded using blockchains. The resource management shortcoming of distributed computational resources, such as NFV, data center, fog nodes at edges in fog computing are energy, storage, and computational resources With the increase of massive IoT, the 5G and beyond networks will employ more constraints in their resources. 5G mainly has three service use cases: Ultra-reliable and low latency (URLLC), massive machine type computing (mMTC), and enhanced mobile broadband (eMMB). The sixth generation (6G) is a revolutionary initiative in the history of wireless networks and promises to support a wide geographical region with ultra-high data rates, massive enabled IoTs, connected drones, virtual reality, and network autonomy (by leveraging the machine learning components in the pipeline) [50]. SDN softwarization, Southbound interface, East/West interface, and on-the-fly management aspects can support the 5G and 6G networks, a wide area of applications, and massive IoTs across the globe. Machine learning, SDNs, and edge cloud are key components of the next generation of cellular networks. 5G relies on edge cloud deployments to satisfy the ultra-low latency demand of future applications. Such deployments are used to enable advanced data-driven and machine learning (ML) applications in mobile networks. A (new) edge-controller-based architecture for cellular networks is proposed to evaluate its performance with real data traffic.

The aspect of resource-limited nodes of edge networks requires efficient resource provisioning in the network for QoS. Phan et al. [51] proposed a dynamic job-offloading mechanism among resource constraints fog nodes. To enable intelligent offloading for appropriate fog nodes, an SDN controller is utilized that can support the offloaded task at minimum cost. The controller can dynamically investigate the resource capabilities, link congestion, and network statistical log files. Using computational offloading among fog nodes decreases the end-to-end latency, traffic detouring to oblivious links, and fog computational resources. A software defined network function virtualization (SDNFV) network utilizes stateful firewall services which are deployed as Virtual network functions (VNFs) to increase network performance, security, and scalability [52]. It utilizes ML algorithms to identify potentially malicious linkages and probable attack targets.

To handle the TCP SYN attack, FUPE (a security-aware task scheduler) is proposed that handles DDoS attacks in a distributed environment [53]. It integrates SDN in its architecture for security objectives. FUPE implements the security-aware task scheduling at the fog gateway. FUPE amalgamates the multi-objective particle swarm optimization algorithm and fuzzy logic for security enhancement. The SDN central resource management unit helps the FUPE with instantaneous decisions in IoT-fog networks. To maintain the security status, FUPE assigns the trusted user application tasks to the trustworthy fog computational devices in its scheduler architecture. Blockchain helps to identify informational alteration at any stage when transactions or information are already preserved in the form of linked blocks. It proposes the modified blockchain leveraged with the SDN controller. The SDN controller helps to register the devices in each domain where the registration information cannot be changed. The SDN controller maintains the public blocks for the registration of devices while the architecture maintains a private blockchain mechanism at device-level communication. Each controller is assisted with a blockchain and storage to keep a record of distributed ledgers. Therefore, public and private key-based domain identification of devices supports the inter-domain mobility, security, and energy-consumption-aware communication in cyber-physical systems (CPS) [54].

It is hard to tackle cyber attacks using traditional security mechanisms. Traditional network equipment and network functions cannot support an efficient defense against the attacks because of the network function's rigidity. SDN controllers support the programable defense applications in various centralized and decentralized networks and can gauge the congestion, port, flow rule entry, and the attached end-user device behavior. SDN employs the detection, localization, and proactive, reactive mitigations against cyber-attacks. To ensure security, softwarized control functions of different domains can collaborate and defend against cyber attacks. DHCPguard [55] exploited the DHCP attacks in networks and provided mechanisms to defend against the attacks. It handles the attacks by utilizing an SDN controller, specifically a security module on top of the POX controller, which is designed for mitigation of DHCP starvation attacks. DHCPguard also facilitates IP pool recovery, DHCP server availability, snooping, and rate-limiting. A traditional DHCP mechanism lacks a security mechanism (i.e., discovery flooding message of the DHCP client program). Compared to the traditional network forwarding devices, SDN architecture can decide the DHCP client application messages at the central controller and block the suspected or malicious nodes at the forwarding devices.

In [56], security assurance is guaranteed through the protocol dialects extension. The protocol directs and carries the objectives to provide robustness against downgrade attacks and facilitates the network protocol in the context of network security. The OFP dialects have derivatives of MAC-based authentication and complete production packet security without message modifications. The packet forwarding optimal decisions in fog computing needs to reconsider periodically for efficient network management. Trust and security parameters in fog computing need to be reconsidered in future networks. A malicious fog node can have forged links with other fog nodes and suspicious activities in production packets, opening the possibility for this sort of node to lock the resources of connected services or alter the topology view at the central SDN controller. Therefore, the resultant computation and energy consumption ratio of fog computing infrastructure also increases [57]. The current SDN standardization, especially in the form of OpenFlow, needs to be upgraded for the flawless integration of fog resources. Fog computing represents a geographical distribution of the resources and host applications, which makes the network more vulnerable. To tackle this vulnerability of the fog paradigm, the distributed architecture of the SDN needs to be more defensive [58].

Vehicular ad hoc networks (VANETs) in integration with SDN supports efficient resource management for the computational offloading of the moving vehicles [59]. Likewise, this is an edge for the various network security issues. VANETs must confront such attacks as man-in-the-middle, DDoS, and jamming. If the SDN layer is vulnerable, then the SDN-enabled VANETs are more complex and less equipped with a defense mechanism. Moving vehicle applications always trust the nearest RSU for computational offloading. In cases of information fabrication or privacy leaks, the end-user trust is declined for the infrastructure. If the central single control functions are under various attack conditions, the SDN forwarding devices in VANETs cannot defend and classify the malicious activities or malicious hosts [60]. In IoT-enabled healthcare infrastructures, device authentication is important. Traditional network devices are unable to authenticate or bootstrap the fresh connected end devices securely. SDN-supported authentication, routing from end device to edge server, and inter-edge server communication (routing for the load balance) need to be critically are analyzed. The proposed framework has an IoT device authentication method that is supported by the probabilistic k-nearest neighbor (p-KNN). The framework uses the p-KNN to evaluate the validity of end IoT devices residing in a healthcare system. Using p-KNN, an edge server investigates the legitimacy of healthcare IoT devices and SDN to performs efficient collaboration among the edge servers that are close computation resources [61].

To localize the DDoS, a convolution neural network (CNN) can be utilized. CNN can perform better for DDoS detection as compared to logistic regression, multi-layered perceptron, and the dense multi-layered perceptron. It uses Game Theory to drop the malicious activities. As a result, it saves the central SDN controller deployed in any ISP from the IoT devices intended for DDoS. If the IoT devices impairs the central controller, then the vulnerability degree of SDN-managed ISP is increased [62]. SDN security issues need serious investigation to overcome DDoS attacks in the controller and communication [63]. It also integrates the online learning method to limit the PacketIn rate while caring for the controller queue and switch space capacity. Traditionally, there are straightforward approaches to limit packet rate, but these cannot be reliable for bandwidth-sensitive applications in a real network. The proposed parallel online deep learning (PODL) algorithm envisions the two important aspects and adjusts the weight for enqueuing the controller packets (PacketIn) and flow rule installation capacity in forwarding devices. In SDN, the controller handles communication traffic and installs flow rules at the switches based on network policies and applications. As the controller is powerful with better memory and processing power, the increase in network traffic growth can be handled and multiple controllers can be installed to avoid overloading and fault tolerance.

Energy conservation is one of the serious problems faced by WSN, as the sensor nodes have limited battery power and are expected to perform data aggregation and actuation functions in addition to sensing data. With the recent technology that introduces network programmability, network trafficking is a prominent domain for the SDN applicability. An network with no infrastructure consisting of self-powered and self-configuring sensor nodes that collect the sensed data and perform aggregation and actuation can function reliably, efficiently, and accurately. To cover a large geographical area and have good connectivity, a sensor network demands a high number of sensor nodes compared to many wireless nodes in a typical ad hoc network [64]. An industrial wireless sensor network (IWSN) is typically installed in a hostile or unmanaged environment where data aggregation is becoming increasingly difficult. Currently, data aggregation techniques are mostly concerned with increasing the efficiency of data transmission and aggregation, with the goal to improve data security. The secure data aggregation techniques' performance is the result of a trade-off between multiple parameters, including transmission/fusion, energy, and the environment. The analysis of the security needs and approaches for WSN data aggregation and provides a comprehensive explanation of the traditional secure data aggregation procedures is presented in [65] It qualitatively assesses the advantages and disadvantages of existing security systems in each category and concludes that security and energy efficiency are suitable for IWSN.

SDN-(UAV) ISE [66] was introduced for WSN with data mules in which sensors can communicate directly to a control center through the long-range wireless interface or by utilizing the multi-hop paradigm via a short-range wireless communication interface. A drone acts as a mobile sink for the latter case. The movement of the data mule is forecast by the SDN controller and the forecasted positions are considered to generate the flow table entries and schedule their applications. WSNs play a fundamental role in environmental monitoring and have been the subject of extensive research efforts for two decades. Recently, several environmental monitoring solutions have been proposed that exploit unmanned aerial vehicles (UAV). WSN nodes are usually deployed in large numbers in the monitored environment and are equipped with low-cost sensors (while drones are less numerous due to their higher cost, they can be equipped with several expensive sensors). A new industrial WSN in the industry that is based on fog computing is the fog industrial wireless sensor network (F-IWSN). It not only minimizes information transmission latency more effectively but also performs real-time control and effective resource allocation. A complete trust management system for F-IWSN based on Gaussian distributions is presented in [67]. In this proposed system, a gray decision-making strategy is used in its trust decision to create a balance of security, transmission performance, and energy consumption. The suggested trade-off, namely a trust management-based secure routing strategy, may successfully pick the secure and robust relay node. Furthermore, the presented strategies can be used to fight against badmouthing assaults. The results reveal that GDTMS outperforms other similar algorithms in terms of overall performance. It prevents network holes from appearing, balances network traffic, and increases network survivability. All these capabilities make SDN a suitable choice in wireless networks.

8.13 CONCLUSION

SDN is an innovative network paradigm that brings simplicity to networks by simplifying network management and automating configuration. A WSN provides a platform for low-rate wireless personal area networks with few resources and short communication ranges. However, it incurs several challenges with respect to network management, security, routing, etc. The SDWSN model helps to resolve these challenges by handling network management and control from the central controller. This chapter comprises network control and management activities regarding SDN-based WSNs. It also discussed SDN methods used in WSNs, which provide flexibility, enhance management, and improve overall performance. Moreover, different network policies are also discussed for SDN-based WSNs. In addition, network policies are discussed with respect to the automation and optimization for effective communication in WSNs. Finally, a comprehensive discussion on different studies is presented, which made SDN a data-driven solution for wireless networks.

REFERENCES

1. N. Feamster, J. Rexford, and E. Zegura, "The road to SDN: An intellectual history of programmable networks," ACM SIGCOMM Computer Communication Review, vol. 44, no. 2, pp. 87–98, Apr. 2014.
2. N. Mckeown, "How SDN will Shape Networking," October 2011, [Online]. Available: http://www.youtube.com/watch?v=c9-K5OqYgA
3. R. Alvigzu, G. Maier, N. Kukreja, A. Pattavina, R. Morro, A. Capello, and C. Cavazzoni, "Comprehensive survey on SDN: Software defined networking for transport networks," IEEE Communications Surveys & Tutorials, vol. 19, no. 4, pp. 2232–2283, 2017.
4. M. Mousa, A. M. Bahaa-Eldin, and M. Sobh, "Software defined networking concepts and challenges," In Proc. Int. Conf. on IEEE Computer Engineering & Systems (ICCES), pp. 79–90, 2016.
5. Z. Latif, K. Sharif, F. Li, M.M. Karim, S. Biswas, and Y. Wang, "A comprehensive survey of interface protocols for software defined networks," Journal of Network and Computer Applications, vol. 156, pp. 102563, 2020.

6. T. Luo, H. P. Tan, and T. Q. S. Quek, "Sensor openflow: Enabling software-defined wireless sensor networks," IEEE Communications Letters, vol. 16, no. 11, pp. 1896–1899, Nov. 2012.

7. S. Costanzo, L. Galluccio, G. Morabito, and S. Palazzo, "Software defined wireless networks: Unbridling SDNS", In European Workshop on Software Defined Networking, pp. 1–6, 2012.

8. L. Galluccio, S. Milardo, G. Morabito, and S. Palazzo, "SDN-wise: Design, prototyping and experimentation of a stateful SDN solution for wireless sensor networks," In IEEE Conference on Computer Communications, INFOCOM, pp. 513–521, Apr. 2015.

9. P. Berde, et al., "ONOS: Towards an open, distributed SDN OS," Proceedings of the Third Workshop on Hot Topics in Software Defined Networking, ACM, 2014.

10. P. Newman, G. Minshall, and T. L. Lyon, "IP switching-ATM under IP," IEEE/ACM Trans. Netw., vol. 6, no. 2, pp. 117–129, April 1998.

11. N. Gude, T. Koponen, J. Pettit, B. Pfaff, M. Casado, N. McKeown, and S. Shenker, "NOX: Towards an operating system for networks," Computer Communication Review, vol. 38, no. 3, pp. 105–110, 2008.

12. OME Committee, "Software-defined networking: The new norm for networks," Open Networking Foundation, 2012.

13. S. Saha, S. Prabhu, and P. Madhusudan, "NetGen: Synthesizing data plane configurations for network policies," Proceedings of the 1st ACM SIGCOMM Symposium on Software Defined Networking Research. ACM, 2015.

14. V. Varadharajan, K. Karmakar, U. Tupakula, and M. Hitchens, "A policy-based security architecture for software-defined networks," IEEE Transactions on Information Forensics and Security, vol. 14, no. 4, pp. 897–912, 2019.

15. T. Lian, Y. Zhou, X. Wang, N. Cheng, and N. Lu, "Predictive task migration modeling in software defined vehicular networks," 4th IEEE International Conference on Computer and Communication Systems (ICCCS), pp. 570–574, 2019.

16. N. Kato, B. Mao, F. Tang, Y. Kawamoto, and J. Liu, "Ten challenges in advancing machine learning technologies toward 6G," IEEE Wireless Communications, 2020.

17. T. Benson, A. Akella, and D. Maltz, "Unraveling the complexity of network management," 6th USENIX Symposium on Networked Systems Design and Implementation, Berkeley, CA, USA, pp. 335–348, 2009.

18. D. L. Tennenhouse, J. M. Smith, W. D. Sincoskie, D. J. Wetherall, and G. J. Minden, "A survey of active network research," IEEE Communications Magazine, vol. 35, no. 1, pp. 80–86, 1997.

19. A. T. Campbell, H. G. De Meer, M. E. Kounavis, K. Miki, J. B. Vicente, and D. Villela, "A survey of programmable networks," SIGCOMM Computer Communication Review, vol. 29, no. 2, pp. 7–23, Apr. 1999.

20. J. Rexford, A. Greenberg, G. Hjalmtysson, D.A. Maltz, A. Myers, G. Xie, J. Zhan, and H. Zhang, "Network-wide decision making: Toward a wafer-thin control plane," In Proc. HotNets, pp. 59–64. Citeseer, 2004.

21. A. Greenberg, G. Hjalmtysson, D. A. Maltz, A. Myers, J. Rexford, G. Xie, H. Yan, J. Zhan, and H. Zhang, "A clean slate 4D approach to network control and management," ACM SIGCOMM Computer Communication Review, vol. 35, no. 5, pp. 41–54, 2005.

22. M. Caesar, D. Caldwell, N. Feamster, J. Rexford, A. Shaikh, and J. van der Merwe, "Design and implementation of a routing control platform," In Proceedings of the 2nd conference on Symposium on Networked Systems Design & Implementation, vol. 2, pp. 15–28. USENIX Association, 2005.

23. M. Casado, M. J. Freedman, J. Pettit, J. Luo, N. McKeown, and S. Shenker, "Ethane: Taking control of the enterprise," ACM SIGCOMM Computer Communication Review, vol. 37, no. 4, pp. 1–12, 2007.

24. Open networking foundation. https://www.opennetworking.org/about. Accessed on 12 March 2022
25. N. McKeown, T. Anderson, H. Balakrishnan, G. Parulkar, L. Peterson, J. Rexford, S. Shenker, and J. Turner, "OpenFlow: Enabling innovation in campus networks," ACM SIGCOMM Computer Communication Review, vol. 38, no. 2, pp. 69–74, 2008.
26. Open Networking Research Center (ONRC). http://onrc.net. Accessed on 12 March 2022
27. J. Yick, B. Mukherjee, and D. Ghosal, "Wireless sensor network survey," Computer Networks, vol. 52, no. 12, pp. 2292–2330, 2008.
28. D. Sharma, S. Verma, and K. Sharma, "Network topologies in wireless sensor networks: A review," International Journal of Electronics & Communication Technology, vol. 4, pp. 93–97, 2013.
29. M. K. Singh, et al., "A survey of wireless sensor network and its types," 2018 International Conference on Advances in Computing, Communication Control and Networking (ICACCCN). IEEE, 2018.
30. N. Xu, "A survey of sensor network applications," IEEE Communications Magazine, vol. 40, no. 8, 102–114, 2002.
31. M. Ndiaye, G. P. Hancke, and A. M. Abu-Mahfouz, "Software defined networking for improved wireless sensor network management: A survey," Sensors, vol. 17, no. 5, pp. 1031, 2017.
32. T. Tony and L. Hiryanto, "Software-Defi wireless sensor networks: A systematic review, architecture and challenges," In IOP Conference Series: Materials Science and Engineering. IOP Publishing, 2020.
33. S. S. G Shiny, S. S. Priya, and K. Murugan, "Repeated game theory-based reducer selection strategy for energy management in SDWSN," Computer Networks, vol. 193, pp. 108094, 2021.
34. J. Ramprasath and V. Seethalakshmi, "Improved network monitoring using software-defined networking for DDoS detection and mitigation evaluation," Wireless Personal Communications, vol. 116, no. 3, pp. 2743–2757, 2021.
35. J. Maruthupandi, et al., "Route manipulation aware Software-Defined Networks for effective routing in SDN controlled MANET by Disney Routing Protocol," Microprocessors and Microsystems, vol. 80, pp. 103401, 2021.
36. U.A. Bukar, and M. Othman, "Architectural design, improvement, and challenges of distributed software-defined wireless sensor networks," Wireless Personal Communications, vol. 122, no. 3, pp. 2395–2439, 2022.
37. S. Manikandan, and M. Chinnadurai, "Effective energy adaptive and consumption in wireless sensor network using distributed source coding and sampling techniques," Wireless Personal Communications, vol. 118, no. 2, pp. 1393–1404, 2021.
38. M. Ibrar, L. Wang, G.-M. Muntean, A. Akbar, N. Shah, and K. R. Malik, "PrePass-Flow: A machine learning based technique to minimize ACL policy violation due to links failure in hybrid SDN," Computer Networks, vol. 184, pp. 107706, Jan. 2021, doi: 10.1016/j.comnet.2020.107706
39. D. W. Frye, 2007. Network Security Policies and Procedures, (vol. 32). Springer Science & Business Media.
40. D. Bertsimas and V. Goyal, "On the power and limitations of affine policies in two-stage adaptive optimization," Mathematical Programming, vol. 134, no. 2, 491–531, 2012.
41. P. J. Goulart, E. C. Kerrigan, and J. M. Maciejowski, "Optimization over state feedback policies for robust control with constraints," Automatica, vol. 42.4, pp. 523–533, 2006.
42. C. Prakash, et al., "PGA: Using graphs to express and automatically reconcile network policies," ACM SIGCOMM Computer Communication Review, vol. 45.4, pp. 29–42, 2015.

43. N. Handigol, et al., "Aster* x: Load-balancing web traffic over wide-area networks," In Proceedings of GENI Engineering Conf, vol. 9. 2010.

44. F. Valenza, S. Spinoso, and R. Sisto, "Formally specifying and checking policies and anomalies in service function chaining," Journal of Network and Computer Applications, vol. 146, no. 102419, Nov. 2019.

45. E. Al-Shaer and S. Al-Haj, "FlowChecker: Configuration analysis and verification of federated OpenFlow infrastructures," Proceedings of the 3rd ACM Workshop on Assurable and Usable Security Configuration. ACM, 2010.

46. H. Mai, et al., "Debugging the data plane with anteater," ACM SIGCOMM Computer Communication Review. vol. 41. no. 4. ACM, 2011.

47. B. Almadani, A. Beg, and A. Mahmoud, "DSF: A distributed SDN control plane framework for the east/west interface," IEEE Access, vol. 9, pp. 26735–26754, 2021.

48. A. Rahman, M. J. Islam, A. Montieri, M. K. Nasir, M. M. Reza, S. S. Band, and A. Mosavi, "SmartBlock-SDN: An Optimized Blockchain-SDN Framework for Resource Management in IoT," IEEE Access, vol. 9, pp. 28361–28376, 2021.

49. Derhab, A., Guerroumi, M., Belaoued, M., & Cheikhrouhou, O. "BMC-SDN: block-chain-based multicontroller architecture for secure software-defined networks," Wireless Communications and Mobile Computing, 2021.

50. Malik, U. M., Javed, M. A., Zeadally, S., & ul Islam, S. "Energy efficient fog computing for 6G enabled massive IoT: Recent trends and future opportunities," IEEE Internet of Things Journal, no. 9984666, 2021.

51. L. A. Phan, D. T. Nguyen, M. Lee, D. H. Park, and T. Kim, "Dynamic fog-to-fog offloading in SDN-based fog computing systems," Future Generation Computer Systems, vol. 117, pp. 486–497, 2021.

52. S. Prabakaran, et al., "Predicting attack pattern via machine learning by exploiting stateful firewall as virtual network function in an SDN network," Sensors, vol. 22.3, pp. 709, 2022.

53. S. Javanmardi, M. Shojafar, R. Mohammadi, A. Nazari, V. Persico, and A. Pescapè, "FUPE: A security driven task scheduling approach for SDN-based IoT–Fog networks," Journal of Information Security and Applications, vol. 60, pp. 102853, 2021.

54. S. A. Latif, F. B. X. Wen, C. Iwendi, F. W. Li-li, S. M. Mohsin, Z. Han, and S. S. Band, "AI-empowered, blockchain and SDN integrated security architecture for IoT network of cyber physical systems," Computer Communications, vol. 181, pp. 274–283, 2022.

55. M. S. Tok, & M. Demirci, "Security analysis of SDN controller-based DHCP services and attack mitigation with DHCPguard," Computers & Security, vol. 109, pp. 102394, 2021.

56. M. Sjoholmsierchio, B. Hale, D. Lukaszewski, and G. Xie, "Strengthening SDN security: Protocol dialecting and downgrade attacks," IEEE 7th International Conference on Network Softwarization (NetSoft), pp. 321–329, 2021.

57. A. Alamer, "Security and privacy-awareness in a software-defined Fog computing network for the Internet of Things," Optical Switching and Networking, pp. 100616, 2021.

58. E. Ahvar, S. Ahvar, S. M. Raza, J. Manuel Sanchez Vilchez and G. M. Lee, "Next generation of SDN in cloud-fog for 5G and beyond-enabled applications: opportunities and challenges," Network, vol. 1, no. 1, pp. 28–49, 2021.

59. R. Sultana, J. Grover, and M. Tripathi, "Security of SDN-based vehicular ad hoc networks: State-of-the-art and challenges," Vehicular Communications, vol. 27, pp. 100284, 2021.

60. W. B. Jaballah, M. Conti, and C. Lal, "Security and design requirements for software-defined VANETs," Computer Networks, vol. 169, pp. 107099, 2020.

61. J. Li, J. Cai, F. Khan, A. U. Rehman, V. Balasubramaniam, J. Sun, and P. Venu, "A secured framework for SDN-based edge computing in IOT-enabled healthcare system," IEEE Access, vol. 8, pp. 135479–135490, 2020.

62. M. V. deAssis, L. F. Carvalho, J. J. Rodrigues, J. Lloret, and M. L. Proença Jr., "Near real-time security system applied to SDN environments in IoT networks using convolutional neural network," Computers & Electrical Engineering, vol. 86, pp. 106738, 2020.

63. A. ElKamel, H. Eltaief, and H. Youssef, "On-the-fly (D) DoS attack mitigation in SDN using deep neural network-based rate limiting," Computer Communications, vol. 182, pp. 153–169, 2022.

64. P. Jayashree and F. Infant Princy, "Leveraging SDN to conserve energy in WSN-An analysis," IEEE 3rd International Conference on Signal Processing, Communication and Networking (ICSCN), 2015.

65. W. Fang, et al., "Comprehensive analysis of secure data aggregation scheme for industrial wireless sensor network," Computers, Materials and Continua, vol. 61.2, pp. 583–599, 2019.

66. J. S. Mertens, et al., "SDN-(UAV) ISE: Applying software defined networking to wireless sensor networks with data mules," IEEE 21st International Symposium on "A World of Wireless, Mobile and Multimedia Networks" (WoWMoM), 2020.

67. W. Fang, et al., "TMSRS: Trust management-based secure routing scheme in industrial wireless sensor network with fog computing," Wireless Networks, vol. 26.5, pp. 3169–3182, 2020.

9 Data-Driven Coexistence in Next-Generation Heterogeneous Cellular Networks

Salman Saadat
Systems Engineering Department, Military
Technological College, Oman

CONTENTS

9.1 INTRODUCTION

The traffic volume per subscriber has seen rapid growth over the last decade due to the proliferation of mobile data applications and devices. With the introduction of more advanced mobile devices and real-time services, such as interactive gaming and video streaming, mobile broadband data usage is expected to grow by 1,000-fold [1]. The dramatic increase in data traffic over cellular networks has posed stiff challenges for mobile network operators since it is hard to accommodate such huge volumes of data with the limited licensed spectrum. Heterogeneous network (HetNet) architecture for radio access networks (RANs) using small cells has, therefore, been adopted for long-term evolution (LTE) operation by 3GPP [2]. Small-cell deployment in HetNet is seen as one of the key techniques to improve the capacity and coverage of future cellular networks. The small cells in HetNet can be deployed to share the spectrum with macro-cells (in-band) or can operate on a different frequency band (out-of-band).

With in-band HetNet deployment, small cells improve coverage and also enhance the spectral utilization by sharing bandwidth with the macro-cell base station (MBS)

DOI: 10.1201/9781003216971-12

by reusing it multiple times inside the macro-cell coverage. However, due to macro-cell and small-cell operation in the same band, strong interference exists between the two tiers. Several studies have shown that the throughput over point-to-point links is approaching its theoretical limit in traditional HetNet. Therefore, to satisfy the growing demand for wireless broadband access to the Internet over the cellular network, out-of-band small-cell deployment using unlicensed bands is considered for LTE operation by 3GPP in LTE Release 13. The use of unlicensed bands offers vast potential for enhancing the capacity of the cellular network by integrating the high spectral efficiency of LTE and the large underutilized bandwidth of the unlicensed spectrum.

For the out-of-band small-cell deployment in next-generation HetNet, the focus has been on unlicensed bands in the 5 GHz and mmWave frequencies. Various strategies for LTE traffic offloading on the unlicensed spectrum have been explored in the literature, including integrated femto-Wi-Fi (IFW), where LTE traffic is offloaded on to Wi-Fi. In IFW implementation, Wi-Fi and cellular interfaces in user devices connect with the Wi-Fi and LTE networks, respectively, for better throughput. However, with licens-assisted access (LAA), the Wi-Fi and cellular spectrum resources are integrated and the user accesses both licensed and unlicensed bands through a single LTE interface for seamless handover and improved quality of service (QoS) management.

LAA uses the carrier aggregation (CA) technique to jointly utilize the licensed and unlicensed channels for realizing high data rates. With LAA, LTE signals are transmitted normally over the licensed carriers, whereas the unlicensed channel is used opportunistically. For implementation of LAA, traffic over the control and data planes is separated. With LAA, the licensed carrier is used for the control plane traffic, while the traffic in the data plane can also utilize unlicensed carriers either as a supplement downlink (SDL) or time division duplex (TDD) channel. Since the regulations on the use of 5 GHz unlicensed bands restricts the maximum transmission power, LAA is feasible only for the small-cell deployment, such as femtocell.

LTE operation over unlicensed bands is a concern for legacy wireless networks using the unlicensed band. The major cause of this concern is the difference in the spectrum access mechanisms of LTE and legacy networks. Due to these concerns, it is significant to develop a fairness aware coexistence scheme for the two systems. The resource allocation problem is especially complicated in case of HetNet deployment due to the large diversity in type of nodes, their service requirements, transmit power, and their unplanned incremental deployment. In this chapter, we will focus on spectrum sharing and coexistence between LAA and Wi-Fi systems, as Wi-Fi is the major incumbent system in the 5 GHz unlicensed band. The LAA-Wi-Fi deployment and contention scenario are described in Figure 9.1.

9.2 SPECTRUM ACCESS ON UNLICENSED BANDS

Despite numerous advantages of using the unlicensed spectrum for LTE operation, it has concerns for the legacy networks over unlicensed bands, specifically the Wi-Fi in 5 GHz bands. These concerns mainly stem from the difference in spectrum access procedures of legacy networks and LTE. Currently, the devices operating on the

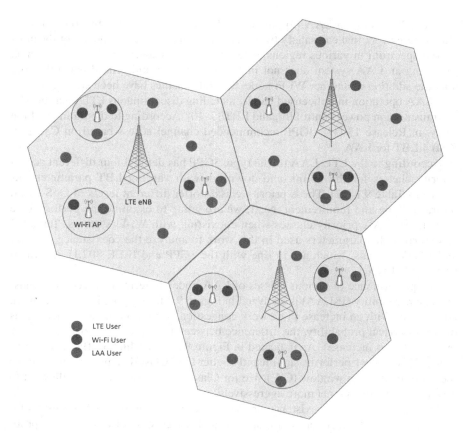

FIGURE 9.1 LAA-Wi-Fi deployment and contention scenario.

5 GHz unlicensed band use contention-based polite spectrum access protocols, such as clear channel assessment (CCA) and carrier sense multiple access with collision avoidance (CSMA/CA), where the channel is sensed for availability before each use and is vacated regularly for the use by other devices operating on the same channel. The spectrum access method of LTE, however, is based on slotted centralized scheduling that does not involve listen, sense, and vacate operations as devices have dedicated access to the licensed channel. Moreover, in CSMA/CA, devices halt transmission and backoff for random amounts of time if they are involved in a packet collision, which is caused by the simultaneous transmissions on a channel. However, in case of LTE, the system responds to packet loss by lowering the transmission rate as it is considered to be the result of channel congestion. In addition, even when LTE systems do not have any data for transmission, they still exchange control traffic at a fixed rate. This means wireless local area network (WLAN) devices, Wi-Fi access points (AP), and stations (STA) will be put in a non-ending backoff by the LTE system if LTE protocols are not modified before their deployment in the unlicensed spectrum. These concerns necessitate changes in the medium access control (MAC) and physical (PHY) layers of LTE for implementation in the 5 GHz unlicensed band.

The 3GPP study on LAA [3] aimed to address these concerns by constituting a global framework that can satisfy the regulatory requirements on the use of the unlicensed spectrum in various regions. It targeted a coexistence mechanism in which deploying an LAA system does not impact the existing unlicensed network more than the addition of another Wi-Fi node. Several schemes have been studied to support LAA operation in unlicensed bands, including discontinuous LTE operation [4, 5], transmission power control [6], and LBT [7–10]. According to the findings of the study, in Release 13 [11], 3GPP recommended channel access based on Category (Cat) 4 LBT for LAA.

According to the LTE-LAA traffic type, 3GPP has defined four different access priority classes for the uplink and downlink with varying LBT parameters, as shown in Table 9.1 [12]. These priority classes offer different levels of QoS based on traffic type and requirement. Next, we are going to examine the performance of different LAA priority classes when coexisting with Wi-Fi systems. Table 9.2 summarizes the parameters used in this work to analyze the coexistence of LAA and Wi-Fi systems, which are in line with the 3GPP and IEEE 802.11 specifications for WLAN.

The performance of various classes of LAA nodes in terms of successful transmission probability (PsLAA) is analyzed in Figure 9.2 for varying levels of network congestion. With an increase in network congestion corresponding to higher values of packet arrival probability, the difference between performance of different LAA priority classes increases. As indicated in Figure 9.2, under high traffic load conditions, LAA Class 1 performs significantly better than Class 4. This is mainly due to the lower contention window (CW) size for Class 1 LAA traffic, which allows it to capture the Wi-Fi channel more aggressively.

The impact of LAA coexistence on Wi-Fi is examined in Figure 9.3. Results for various classes of LAA traffic are studied and the successful transmission probability of Wi-Fi (PsWi-Fi) is analyzed for changing network congestion in terms of packet arrival probability. The performance of the Wi-Fi system varies depending on

TABLE 9.1

Access Priority Classes Based on LTE-LAA Traffic Type

Access Class	CWmin	Backoff Stages	TXOP
Uplink:			
1	4	1	2 msec
2	8	1	3 msec
3	16	2	6–10 msec
4	16	6	6–10 msec
Downlink:			
1	4	1	2 msec
2	8	1	3 msec
3	16	2	8–10 msec
4	16	6	8–10 msec

TABLE 9.2

LAA-Wi-Fi Coexistence Parameters

Definition	Value
Wi-Fi Max. Backoff Stages	5
Wi-Fi Min. Contention Window	16
Number of Bits per Packet	12,000
Channel Rate	300 Mbps
DIFS Duration	34 μs
SIFS Duration	16 μs
Slot Time	9 μs
MAC Header Size	272 bits
PHY Header Size	128 bits
Propagation Delay	0.1 μs
ACK	112 bits + PHY Header

the priority class of the coexisting LAA traffic. The successful transmission probability is lowest for Wi-Fi when coexisting with Class 1 LAA, as the channel is mostly occupied by LAA traffic and Wi-Fi nodes are in constant backoff.

This disparity in performance of the two systems is also highlighted by several studies [13, 14], that indicate that the 3GPP Cat 4 LBT fails to ensure efficient spectrum sharing and fair coexistence, especially under dynamic coexistence

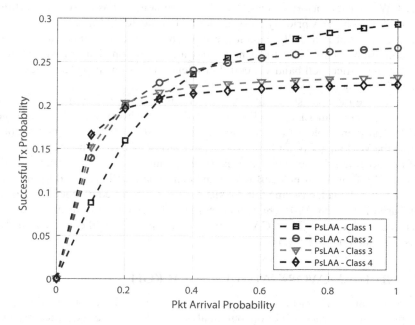

FIGURE 9.2   Successful transmission probability of LAA.

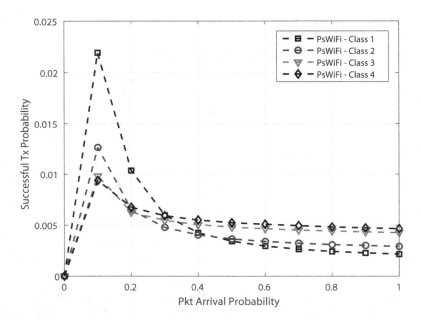

FIGURE 9.3   Successful transmission probability of Wi-Fi.

environments. It is noted that the standard Cat 4 LBT does not consider time-varying Wi-Fi activity over the unlicensed channel as it sets fixed bounds on the LAA CW for each priority class. Moreover, the static transmission opportunity (TXOP) for each LAA priority class, as defined in the 3GPP Cat 4 LBT algorithm, makes it impossible to fully exploit the unlicensed spectrum under low network load conditions. This calls for a new innovative solution that can substantially improve the unlicensed band's spectral efficiency while ensuring fair coexistence of the heterogeneous systems.

To improve the coexistence of LAA and Wi-Fi systems in the unlicensed Wi-Fi band, we propose an LBT-based channel access framework for LAA in the next section, and then explore the machine learning solutions that can exploit available data on the Wi-Fi activity for better spectrum sharing by LAA. The proposed LAA-Wi-Fi coexisting scheme employs backoff on the number of frames transmitted in a single TXOP by LAA eNodeB (eNB) and can dynamically respond to the variations in network load, ensuring fair coexistence in a distributed manner [15, 16]. The LBT based LAA and Wi-Fi channel access procedures on the unlicensed band are described in Figure 9.4. The CCA and transmission process on the unlicensed channel is described in Figure 9.5.

9.3   LBT FRAMEWORK WITH TXOP BACKOFF

We propose an LAA-Wi-Fi coexistence solution based on LBT for downlink LAA. The proposed scheme builds on our analysis of different coexistence solutions studied in the literature while taking into account the coexistence requirements

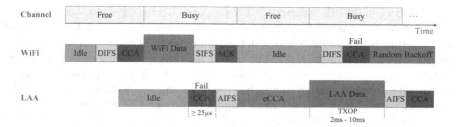

FIGURE 9.4   LAA and Wi-Fi transmission procedure on the unlicensed channel.

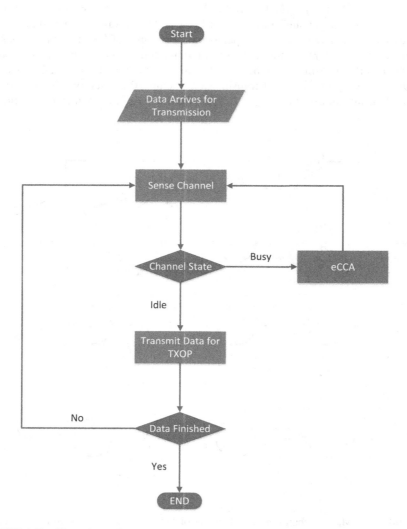

FIGURE 9.5   Clear channel assessment and transmission process.

mentioned in 3GPP LTE Release 13. The proposed LBT coexistence framework for LAA employs adaptive CW size along with exponential backoff on the number of LTE subframes that an LAA eNB can transmit in one TXOP. The backoff on maximum number of subframes per TXOP reduces the channel occupation of LAA as the WLAN load increases, thus implementing better LAA-Wi-Fi coexistence relative to the LBT schemes that ensure high throughput and spectrum efficiency of LTE in its deployment over the unlicensed band just through CW adjustment. The analytical model based on the Markov chain for the proposed LBT scheme with TXOP backoff is presented in Figure 9.6. In detail, the LBT scheme has adaptive CW size, which is updated according to the LAA eNB assessments and user feedback on a comparatively slow time scale. When an LAA eNB needs channel access for packet transmission, it selects a random backoff interval from the CW. In the proposed scheme, the number of subframes that an LAA node can transmit in a single TXOP T is exponential. The value of T depends on the number of failed transmissions for a transmission attempt. At the first transmission attempt, the value of T is set to be T_o, which is the maximum number of subframes that can be transmitted in a single TXOP. For every unsuccessful transmission attempt, the value of T is reduced by half and the minimum value is $T_m = 2^{-m} T_o$. This decreases the channel occupation by

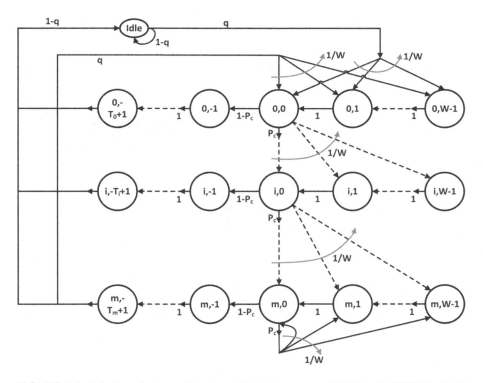

FIGURE 9.6 Markov chain analytical model of the proposed LBT with TXOP backoff scheme for LAA.

LAA with respect to the load over the unlicensed channel, therefore creating more transmission opportunities for the competing nodes. The value of T is restored to T_o after a successful transmission attempt. In Figure 9.6, m is the number of backoff stages, q is the probability of packet arrival for transmission at eNB, and p is the probability of collision. Similar to distributed coordinated function (DCF) protocol for Wi-Fi [17], LAA eNB transmits when the CW backoff counter reaches zero. The CW backoff counter in the proposed LBT framework for LAA decreases when the channel is sensed to be idle, stays the same when the channel is sensed to be busy, and decreases again if the channel is sensed idle for a duration that is larger than distributed inter-frame space (DIFS). The Markov chain model considered for Wi-Fi in this study is based on Bianchi's work [17] and has exponential backoff CW size, which doubles for each transmission failure and resets to minimum CW value for every successful transmission. In general, the Markov model considered here is characterized by an idle state with probability π_I, different backoff states with probability $\pi_{m,W}$, and transmission states with probability $\pi_{m,-T}$. For backoff states, W represents the backoff window. For transmission states, T is the number of subframes in an TXOP. In this chapter, we concentrate on the successful channel access probability of each system as the corresponding load increases. Moreover, the maximum number of subframes in an TXOP is normalized to 1 for analysis and comparison. To evaluate the performance, we compare the proposed scheme with the traditional schemes proposed in literature that do not support backoff on frames per TXOP.

The closed form solution for LAA Markov chain with TXOP backoff and probability of packet arrival can be presented as

$$
\pi_{i,k} =
\begin{cases}
p^i \pi_{0,0} & 0 < i < m, k = 0 \\[2mm]
\dfrac{p^m}{1-p} \pi_{0,0} & i = m, k = 0 \\[2mm]
\dfrac{W-k}{W} q\kappa & i = 0, k\epsilon(1, W-1) \\[2mm]
\dfrac{W-k}{W} p\pi_{i-1,0} & 0 < i < m, k\epsilon(1, W-1) \\[2mm]
\dfrac{W-k}{W} p(\pi_{m-1,0} + \pi_{m,0}) & i = m, k\epsilon(1, W-1) \\[2mm]
(1-p)p^i \pi_{0,0} & i\epsilon(0,m), k\epsilon(0, -T_i + 1)
\end{cases}
\qquad (9.1)
$$

$$
\pi_I = (1-q)\pi_I + (1-q)(1-p)\sum_{j=0}^{m} \pi_{j,-T_j+1}, \qquad (9.2)
$$

where p is the probability that the transmitted packet will encounter a collision and κ is given by

$$
\kappa = (1-p)\sum_{j=0}^{m} \pi_{j,-T_j+1} + \pi_I. \qquad (9.3)
$$

The above boundary conditions can be further simplified by expressing them in terms of $\pi_{0,0}$ and making use of the facts that, $p\pi_{i-1,0} = \pi_{i,0}$, $\pi_{i,0} = p^i\pi_{0,0}$ *for* $0 < i < m$, and $\pi_{j,-T_j+1} = \pi_{j,0}$. Also as $p\pi_{m-1,0} = (1-p)\pi_{m,0}$, $\pi_{m,0} = \frac{p^m}{1-p}\pi_{0,0}$. Moreover,

$$\sum_{j=0}^{m}\pi_{j,-T_j+1} = \sum_{j=0}^{m-1}\pi_{j,-T_j+1} + \pi_{m,-T_m+1}$$

$$= \pi_{0,0}\left(\frac{1}{1-p}\right). \tag{9.4}$$

Therefore, the simplified closed form solution for the Markov chain can be expressed as

$$\pi_{i,k} = \begin{cases} p^i\pi_{0,0} & i\epsilon(0,m-1), k=0 \\[2mm] \dfrac{p^m}{1-p}\pi_{0,0} & i=m, k=0 \\[2mm] \dfrac{W-k}{W}q(\pi_{0,0}+\pi_I) & i=0, k\epsilon(0,W-1) \\[2mm] \dfrac{W-k}{W}\pi_{i,0} & 0<i\leq m, k\epsilon(0,W-1) \\[2mm] (1-p)p^i\pi_{0,0} & i\epsilon(0,m), k\epsilon(0,-T_i+1) \end{cases} \tag{9.5}$$

$$\pi_I = (1-q)\pi_I + (1-q)\pi_{0,0} = (1-q)(\pi_I + \pi_{0,0}). \tag{9.6}$$

Applying the normalization condition as follows, we can obtain the value of $\pi_{0,0}$

$$1 = \pi_I + \sum_{i=0}^{m}\sum_{k=0}^{W-1}\pi_{i,k} + \sum_{i=0}^{m}\sum_{k=1}^{T_i-1}\pi_{i,-k}, \tag{9.7}$$

$$\pi_{0,0} = \frac{2q(1-p)\left(1-\dfrac{p}{2}\right)}{\zeta}, \tag{9.8}$$

where ζ is given by

$$\zeta = (1-q)(1-p)\left(1-\frac{p}{2}\right)[2+q(W+1)]$$

$$+(W+1)\left(1-\frac{p}{2}\right)[q^2(1-p)+pq]$$

$$+2q\left[T(1-p)\left(1-\left(\frac{p}{2}\right)^m\right)+\left(T\left(\frac{p}{2}\right)^m-1\right)\left(1-\frac{p}{2}\right)\right]. \tag{9.9}$$

Now, the probability of transmission $P_{t,j}$ of LAA eNB to transmit a packet in any random slot time can be found as

$$P_{t,j} = \sum_{i=1}^{m}\sum_{k=1}^{T_i-1}\pi_{i,-k} + \sum_{i=0}^{m}\pi_{i,0},$$

$$= \pi_{0,0}\left[\frac{\beta}{(1-p)\left(1-\dfrac{p}{2}\right)}\right], \tag{9.10}$$

where β is given by

$$\beta = T(1-p)\left(1-\left(\frac{p}{2}\right)^m\right) + T\left(1-\frac{p}{2}\right)\left(\frac{p}{2}\right)^m. \tag{9.11}$$

Accordingly, using the Markov model presented by Bianchi [17] and considering $\hat{T}$ number of subframes in an TXOP, the probability of transmission $P_{t,i}$ of Wi-Fi to transmit a packet in any random time slot is determined as

$$P_{t,i} = \pi_{0,0}\left(\frac{\hat{T}}{1-\hat{p}}\right), \tag{9.12}$$

where $\hat{p}$ is the probability of collision for Wi-Fi, and $\pi_{0,0}$ for Wi-Fi is given by

$$\pi_{0,0} = \frac{2\hat{q}(1-2\hat{p})(1-\hat{p})}{\xi}, \tag{9.13}$$

where $\hat{q}$ is the probability of packet arrival at Wi-Fi node for transmission and ξ is equal to

$$\xi = 2(1-2\hat{p})(1-\hat{q})(1-\hat{p}) + \hat{q}(1-2\hat{p})(\hat{W}+1)(1-\hat{q})(1-\hat{p})$$

$$+\hat{q}^2(1-2\hat{p})(1-\hat{p})(\hat{W}+1) + \hat{q}\hat{W}(1-\hat{p})\left[2\hat{p}-(2\hat{p})^m\right]$$

$$+\hat{q}(1-2\hat{p})(\hat{p}-\hat{p}^m) + \hat{q}\hat{p}^{\hat{m}}(1-2\hat{p})(2^m\hat{W}+1)$$

$$+2\hat{q}(1-2\hat{p})(\hat{T}-1), \tag{9.14}$$

where $\hat{W}$ is the minimum CW size and $\hat{m}$ is the number of backoff stages for the Wi-Fi.

Coexistence performance of LAA and Wi-Fi systems is evaluated by studying the collision and successful transmission probability for both LAA and Wi-Fi systems under varying network load. The collision probabilities $p_{f,j}$ and $p_{f,i}$ for LAA eNB and Wi-Fi, respectively, that at least two nodes transmit in the same time slot are given as follows

$$p_{f,j} = 1 - \prod_{z \in LAA, z \neq j} \left(1 - P_{t,z}\right) \prod_{i \in WiFi} \left(1 - P_{t,i}\right), \tag{9.15}$$

$$p_{f,i} = 1 - \prod_{j \in LAA} \left(1 - P_{t,j}\right) \prod_{z \in WiFi, z \neq i} \left(1 - P_{t,z}\right). \tag{9.16}$$

Here, $p_{f,j}$ is the collision probability p for LAA eNB j while $p_{f,i}$ is the collision probability $\hat{p}$ for Wi-Fi node i.

Let $P_{tr,j}$ and $P_{tr,i}$ be the transmission probabilities that at least one node in LAA and Wi-Fi systems, respectively, transmits a packet in the considered time slot. These probabilities are represented as follows

$$P_{tr,j} = 1 - \prod_{j \in LAA} \left(1 - P_{t,j}\right), \tag{9.17}$$

$$P_{tr,i} = 1 - \prod_{i \in WiFi} \left(1 - P_{t,i}\right). \tag{9.18}$$

Then, the successful transmission probabilities $P_{s,j}$ and $P_{s,i}$ for LAA eNB and Wi-Fi, respectively, that the considered packet transmission is successful such that only one node transmits while others defer their transmission are presented as follows

$$P_{s,j} = \frac{NP_{t,j} \prod_{z \in LAA, z \neq j} \left(1 - P_{t,z}\right) \prod_{i \in WiFi} \left(1 - P_{t,i}\right)}{P_{tr,j}}, \tag{9.19}$$

$$P_{s,i} = \frac{nP_{t,i} \prod_{z \in WiFi, z \neq i} \left(1 - P_{t,z}\right) \prod_{j \in LAA} \left(1 - P_{t,j}\right)}{P_{tr,i}}, \tag{9.20}$$

where N and n are the total number of LAA and Wi-Fi nodes sharing the unlicensed channel, respectively.

To analyze throughput of the two systems, let Thr_j and Thr_i be the throughput for LAA eNB and Wi-Fi, respectively. Then

$$Thr_j = \frac{P_{tr,j} P_{s,j} \left(1 - P_{tr,i}\right) P_L}{T_{state}}, \tag{9.21}$$

$$Thr_i = \frac{P_{tr,i} P_{s,i} \left(1 - P_{tr,j}\right) P_W}{T_{state}}, \tag{9.22}$$

where P_L is the number of bits in a LAA packet, P_W is the number of bits in a Wi-Fi packet, and T_{state} is the expected time spent per state and can be calculated as described by Song et al. [18]. The value of T_{state} depends on the time duration of three events i.e., expected time of successful transmission for LAA (Wi-Fi) $T_{s,j}$ $(T_{s,i})$, expected time of collision for LAA (Wi-Fi) $T_{c,j}$ $(T_{c,i})$, and expected time of no transmission T_n. The values of $T_{s,i}$, $T_{c,i}$, $T_{c,j}$, and T_n can be calculated according to the method proposed by Bianchi [17]. The values of $T_{s,j}$ can be calculated as follows [19]

$$T_{s,j} = AIFS + TL\left(T_{PHY} + T_{P_L} + 2SIFS + 2\delta + T_{BA}\right) - SIFS, \qquad (9.23)$$

where TL is the average number of frames per TXOP, T_{P_L} is the transmission time of packet payload, δ is the propagation time, $AIFS$ is arbitration inter-frame space, $SIFS$ is time of short interframe space, and T_{BA} is transmission time of block acknowledgement.

The results of successful transmission probability of LAA and Wi-Fi systems with the proposed spectrum access scheme are compared to the traditional approaches in Figure 9.7. Specifically, we select a constant CW scheme for LAA similar to Cat 4 LBT scheme adopted by 3GPP and an adaptive CW scheme with constant TXOP for analysis. The successful transmission probability for the two systems is studied for increasing LAA load over the unlicensed channel. The results indicate that the proposed solution improves the coexistence between the LAA and Wi-Fi systems by increasing the successful transmission probability of Wi-Fi nodes even for high LAA traffic compared to the two other schemes. The improvement in transmission opportunities for the coexisting Wi-Fi nodes under the proposed scheme is result of dynamic adjustment in the channel occupation

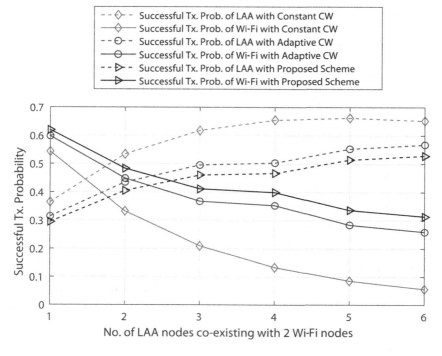

FIGURE 9.7 Successful transmission probability of LAA and Wi-Fi systems for varying LAA load in the network.

period of LAA with the variation in network load. The results also indicate that LAA channel access approach with constant CW gives the lowest successful transmission probability for the coexisting Wi-Fi nodes and the coexistence performance gets even worse with the increase in the number of LAA nodes sharing the channel. With constant CW, LAA nodes cannot respond to the changes in network traffic and keep accessing the unlicensed channel aggressively, leaving no room for transmission by the Wi-Fi.

The results in Figure 9.8 describe the overall throughput improvement on the shared unlicensed channel with the proposed spectrum access scheme for varying network configuration. As the number of Wi-Fi nodes increases on the channel relative to LAA, overall improvement in network throughput decreases since LAA nodes offer better spectrum utilization compared to Wi-Fi due to the dynamic adjustment in CW and varying TXOP based on network congestion. However, the results show that when LAA and Wi-Fi coexist over the unlicensed band, there is a significant improvement in overall network throughput as compared to the all-Wi-Fi network with same number of total nodes. The performance gain of the proposed channel access scheme can be further improved by incorporating machine learning tools to identify the channel access pattern of incumbent devices on the unlicensed band and utilizing this knowledge to improve coexistence between the LAA and Wi-Fi systems.

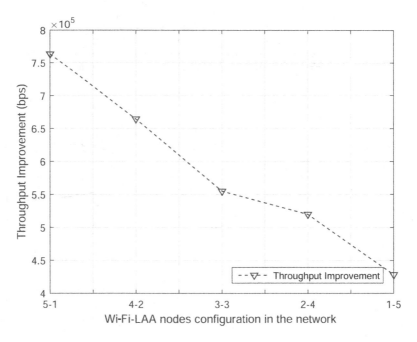

FIGURE 9.8 Throughput-improvement with LAA-Wi-Fi coexistence as compared to an all-Wi-Fi deployment scenario.

9.4 DATA-DRIVEN MACHINE LEARNING APPROACHES TO EFFICIENT RESOURCE SHARING

The resource management in heterogeneous wireless networks with ever-growing diversity and complexity is becoming intractable. This has led to the adoption of emerging machine learning techniques to assist mobile network operators in optimizing network resources and parameters. Machine learning is extensively explored in recent research to enhance the performance of complex heterogeneous networks and is considered a promising tool for resource management in next-generation wireless networks.

Various machine-learning methods and algorithms have been studied in literature for efficient resource sharing by LTE in both the licensed and unlicensed spectrum. A neural network with logistic regression is employed by Yazid et al. [20] to determine performance indicators for LTE-LAA and Wi-Fi links under coexistence. The normalized throughput of the coexisting systems over the unlicensed band is improved under imperfect MAC and PHY layer knowledge using the gradient decent optimization technique over the data set generated for an indoor scenario. The artificial neural network (ANN)-based dynamic spectrum access solution for LTE uplink is discussed by Sahoo [21]. Real-world LTE uplink data are collected from two different locations and a simple perceptron-based ANN is developed to predict transmission opportunities with the duration of previous idle periods on the channel and the required channel occupation time as the input features. A sparse learning framework to eliminate narrowband interference (NBI) to LTE using temporal correlation is discussed by Liu et al. [22]. NBI recovery is formulated as a nonconvex sparse combinatorial optimization problem and probability distribution of NBI is iteratively learnt through machine learning while minimizing the cross-entropy as a loss function.

Supervised machine-learning techniques are explored by Magrin et al. [23] to address collision resolution on the random access channel (RACH) of LTE. The interfering devices that have selected the same preamble are identified at eNB by modeling it as a classification and logistic regression problem. Different machine-learning techniques are tested by Dziedzic et al. [24] to estimate the number of competing Wi-Fi nodes using time series data on Wi-Fi activity. The number of active Wi-Fi APs is determined by observing their energy levels during the LTE off periods and the efficiency of a machine-learning approach is evaluated relative to conventional methods. The saturation level of the Wi-Fi network is classified by Girmay et al. [25] with the help of convolutional neural networks, which is then leveraged for spectrum sharing by LTE. Network statistics on the histogram of inter-frame spacing (IFS), IFS duration, and percentage of frame collision are used to train the machine learning model. LTE network management and service level assurance using the few-shot learning algorithm is examined by Aoki et al. [26]. The semi-supervised prototypical network is considered to reduce the requirement on labeled data to train the classifier model that groups eNB key performance indicators (KPIs) according to the area covered by the eNB. The data set used in the study is gathered from a live LTE network.

A deep reinforcement learning approach with the stochastic gradient descent (SGD) algorithm is adopted for fair spectrum sharing between LAA with Wi-Fi systems by Tan et al. [27]. A deep Q-network (DQN) is built to optimize transmission in each frame by learning traffic patterns of an unsaturated Wi-Fi system without any signaling exchange between the two systems. The data on number of idle slots, number of successfully transmitted Wi-Fi packets, and number of collisions in a frame are used to learn the status of the Wi-Fi network for decision making.

Access class barring is dynamically controlled using reinforcement learning based on dueling Q-Network by Bui and Pham [28] to minimize LTE access delay when coexisting with massive machine-type devices (MTDs). A deep neural network-based scheme is proposed that adjusts both mean barring time and barring factor to optimize access delay and energy consumption. The downlink scheduling problem in LTE is addressed by Robinson and Kunz [29] using deep reinforcement learning. The model takes the channel quality indicator and user equipment's buffer size as input features that are generated through LTE simulation. Network load in a heterogeneous environment is optimized by Stepanov et al. [30] using a support vector machine (SVM) approach. Network edge traffic is predicted using a public cellular traffic data set and compared with bagging and random forest techniques in terms of various quality metrics.

Optimal parameters to accurately predict cellular network performance through a collaborative learning approach are identified by Zeng et al. [31]. A factorization machine-based method is proposed that exploits the correlation between network elements and their historical parameters over time. The results for the average of channel quality indicator (ACQI) and the proportion of multiple-input multiple-output spatial modulation (PMIMOSM) are presented based on the data set collected from a commercial LTE network deployed in a metropolitan area.

Machine-learning solutions are powerful tools to efficiently determine the most suitable operating conditions for a heterogeneous network. However, their performance may vary with changing environments and available training data. Most of the state-of-art research on LAA and Wi-Fi coexistence uses simulation data to train machine learning models, which might not be very accurate in practical scenarios. In the following section, we will analyze the data collected from a live Wi-Fi network and study how it can be utilized along with the machine-learning techniques to improve LAA-Wi-Fi coexistence through improvement in the spectrum sharing procedure.

9.5   CASE STUDY: EXPLOITING DATA ON USER ACTIVITY OVER WI-FI CHANNELS TO IMPROVE CHANNEL SELECTION PROCEDURES

In this section, we explore a public Wi-Fi data set and analyze how the insights gained from the Wi-Fi traffic can assist in better coexistence with LAA systems. The data set contains values of captured Wi-Fi signal in an office location and is based on nine surveys of different lengths [32]. The data set is first employed to determine a pattern of office occupancy by people based on the variations in Wi-Fi signal statistics using various metrics. This information is then exploited for selecting the most suitable Wi-Fi channel for spectrum sharing by LAA.

TABLE 9.3

Features of the Original Data Set

Features	Description	Data Class
Time	Time of sample collection	Object
MAC Address	Layer-2 address of the radio interface	Object
Vendor	Name of the Wi-Fi AP manufacturer	Object
SSID	Network name	Object
Signal Strength	Received signal power in dBm	Float64
Channel	Wi-Fi channel of operation	Float64
Survey	Survey Number of the observation	Float64

The data are collected using the Homedale software running on a computer in an office setup. The data set contains samples collected in the Wi-Fi band every 60 seconds with a total of 648,000 observations. The data collected by this software is not continuous in time and contains gaps caused by various reasons such as computer shutdown/restarts, holidays, and weekends. The features collected in this data set include time, MAC address, vendor name, service set identifier (SSID), signal strength, Wi-Fi channel number, and survey number. The detailed description of these features and their data classes are given in Table 9.3.

The collected data not only contain the signals from the office Wi-Fi Aps, but also captures signals from other office devices operating in the same frequency band, such as printers and personal devices. As indicated in Table 9.3, data contain different data classes, which need to be pre-processed and filtered before analysis. The office Wi-Fi network uses two specific SSIDs, so the data are filtered for these SSIDs for further analysis. Since the same SSID can be broadcast by multiple APs to allow users to roam freely without losing Wi-Fi connectivity, data are therefore grouped by MAC addresses for the selected SSIDs to observe the actual radio connections. The string data in the data set is replaced with integer values through one-hot encoding for easy handling and processing by the machine learning model. The original data set also contained some null entries for certain features. The observations containing these null entries were removed from the data set before analysis. The number of these null entries for different data set features is described in Table 9.4. Moreover, the data set entries corresponding to weekends are not considered in this analysis as they do not reflect the network traffic and congestion experienced on a working day at the office and could possibly bias the data. The number of observations collected for Fridays in the data set were significantly less compared to the other working days and, therefore, were also not considered in the analysis.

We group the Wi-Fi devices by their MAC addresses and then sort them by the Wi-Fi channel they used. It is observed that the office Wi-Fi network mainly uses Wi-Fi channels 1, 6, and 11, which will also be the focus of this study for spectrum sharing with LAA. The number of entries for each Wi-Fi channel in the original data set is given in Table 9.5. In Table 9.6, number of observations are grouped by the channel number and Wi-Fi AP.

TABLE 9.4

Number of Null Entries in the Data Set

Features	Null Entries
Time	0
MAC Address	0
Vendor	234,292
SSID	36,027
Signal Strength	25,183
Channel	25,183
Survey	25,183

TABLE 9.5

Number of Entries for Different Wi-Fi Channels

Channel Number	Number of Entries
1.0	229,630
2.0	231
3.0	2,741
4.0	935
5.0	83
6.0	180,277
8.0	935
9.0	1,660
10.0	5
11.0	206,003
12.0	509

TABLE 9.6

Wi-Fi Channel Used by Different APs in the Office Network

Access Point	Channel Number	Number of Entries
AP1	11.0	59,983
AP2	1.0	53,861
AP3	11.0	24,219
AP4	6.0	19,811

From the analysis of standard deviation and variance of Wi-Fi signal strength at different times of the day, it is observed that higher signal strength variations correspond to higher office occupancy. Moreover, the variation in signal strength depends on population density in the office at any given time. These changes in signal strength are result of people's movements and multipath fading during the office working hours. We can then exploit this variation in signal strength of APs to determine which Wi-Fi channel is more likely to be congested compared to the others at any given time of day. The AP with higher variation in signal strength is considered to have more users within its coverage area, and, therefore, the Wi-Fi channel being used by that AP is more likely to be congested. This insight can then assist in selecting the most appropriate channel for spectrum sharing by LAA.

We analyze the data for most suitable Wi-Fi channel at any given hour of the day that would have a comparatively lower network load relative to the other Wi-Fi channels and then use this channel for LAA. In Figure 9.9, we plot the mean standard deviation (SD) of Wi-Fi signal strength for all office APs for each hour of the day. The values for overall mean SD indicate the variation in Wi-Fi signal strength during the day. These results highlight that the variations in signal strength are highest during office working hours and low during non-business hours, i.e., 6PM to 6AM. This refers to the correlation between the variation in Wi-Fi signal strength and presence of people in the office. The variation in signal strength increases in the morning as the number of people in the office increases and tapers off at the end of the day as people leave the office. We further analyze this trend for individual APs to identify the most appropriate Wi-Fi channel for sharing by LAA.

The SD of signal strength for individual APs is plotted in Figure 9.10. The values of SD indicate the mean variation in signal strength during each hour of the day. Overall, for all APs, the variation in Wi-Fi signal strength increases with the start of office business hours in the morning and decreases with the end of the working hours. We can also observe a deviation from this general trend specifically for AP3. This could be due to the presence of other staff or workers even after working hours

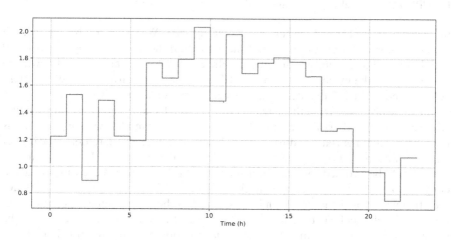

FIGURE 9.9   Overall mean standard deviation of signal strength per hour of all APs.

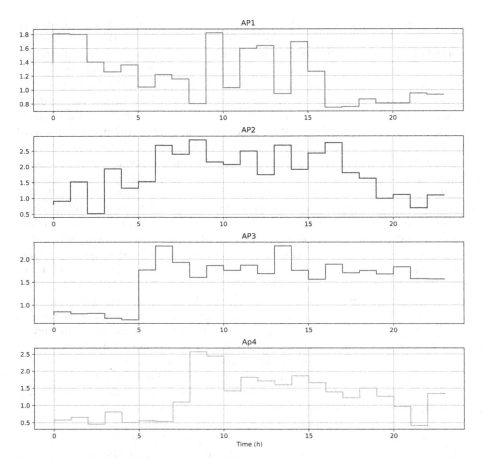

FIGURE 9.10   Signal strength standard deviation variation per hour.

in a certain part of the office that is covered by AP3. We can also observe that the variation in signal strength is highest during certain hours of the day for each AP and these periods of time are different for different APs. We can use this knowledge while training the machine learning model to direct LAA traffic to the Wi-Fi channel assigned to the AP that demonstrates least variation at any given time.

It can be observed from these results that some APs have higher signal variation compared to the others at the same time of day. For example, AP2 in general has the highest signal variations during office business hours, which refers to the higher number of users in its coverage area. Therefore, Wi-Fi channel 1 that is assigned to AP2 should be avoided for spectrum sharing by LAA during these times to avoid congestion and service disruption to the existing Wi-Fi users. Similarly, at 10AM, AP1 that uses Wi-Fi channel 11 in its coverage has the lowest signal strength variation. Wi-Fi channel 11, therefore, can be preferred by LAA when accessing the unlicensed band at this time of day. The information gained from this knowledge on user activity over the Wi-Fi network can contribute significantly to improving the coexistence between the LTE and Wi-Fi networks over the unlicensed band. The channel selection problem for LAA over the unlicensed band can be formulated as a multiclass-classification problem. In the future, we would like to test the prediction accuracy of various machine learning models for different optimization algorithms and loss functions. We would also like to further evaluate and study the coexistence performance of the two systems and improvement in network KPIs when using these models.

9.6 CONCLUSIONS

In this chapter, we have explored the implementation of cellular HetNet in unlicensed frequency bands to address the resource constraints over the licensed spectrum. Specifically, we studied the use of the Wi-Fi band by LTE and addressed the coexistence issues between the Wi-Fi and LTE-LAA systems. It is observed that these coexistence challenges arise mainly due to the difference in channel access mechanism of the two systems. We examined the LBT-based channel access framework adopted by 3GPP for LAA and proposed a channel access solution with adaptive CW and TXOP backoff to further improve the coexistence with Wi-Fi. Data-driven machine-learning solutions to improve coexistence between the two systems are also discussed. We presented a case study where the insight gained from the data collected from a typical Wi-Fi network is exploited to improve the unlicensed channel selection procedure for LAA. The results indicate that there is still a need to improve the LBT procedures and parameters adopted by 3GPP for spectrum sharing by LAA for better coexistence with the incumbent system on the unlicensed spectrum. The results also indicate how data collected from the existing Wi-Fi networks can be exploited along with machine-learning techniques to further improve this coexistence.

REFERENCES

1. L. Nielsen, A. F. Cattoni, A. Diaz Zayas, B. Garcia Garcia, A. Gavras, M. Dieudonne, and E. Kosmatos, "5GPP Whitepaper: Basic testing guide — A starter kit for basic 5G KPIs verification," *Zenodo*, 2021. Available at: https://zenodo.org/record/5704519#. Y1nAeuTMJPY. Accessed on Feb. 24, 2022.
2. J.-S. Liu, C.-H. R. Lin, and Y.-C. Hu, "Joint resource allocation, user association, and power control for 5G LTE-based heterogeneous networks," *IEEE Access*, vol. 8, pp. 122654–122672, 2020.
3. 3GPP.org, "Study on licensed-assisted access using LTE," document RP-141646, RAN meeting #65, Sep. 2014.
4. M. K. Maheshwari, A. Roy, and N. Saxena, "DRX over LAA-LTE-A new design and analysis based on Semi-Markov model," *IEEE Transactions on Mobile Computing*, vol. 18, no. 2, pp. 276–289, 2018.
5. Z. Qin, A. Li, and H. Wang, "A Cournot game approach in the duty cycle in CSAT algorithm for WiFi/LTE-U coexistence," in *International Conference on Wireless Communications and Signal Processing (WCSP)*, pp. 1076–1081, IEEE, 2020.
6. H. Kushwaha, V. Kotagi, and S. R. Murthy, "On the effects of transmit power control on multi carrier LAA-WiFi coexistence," *IEEE Transactions on Sustainable Computing*, vol. 7, no. 3, pp. 656–667, 2021.

7. Q. Ren, J. Zheng, J. Xiao, and Y. Zhang, "Performance analysis of an LAA and WiFi coexistence system using the LAA category-4 LBT procedure with GAP," *IEEE Transactions on Vehicular Technology*, vol. 70, no. 8, pp. 8007–8018, 2021.

8. J. Xiao, J. Zheng, L. Chu, and Q. Ren, "Performance modeling and analysis of the LAA category-4 LBT procedure," *IEEE Transactions on Vehicular Technology*, vol. 68, no. 10, pp. 10045–10055, 2019.

9. J. Zheng, J. Xiao, Q. Ren, and Y. Zhang, "Performance modeling of an LTE LAA and WiFi coexistence system using the LAA category-4 LBT procedure and 802.11e EDCA mechanism," *IEEE Transactions on Vehicular Technology*, vol. 69, no. 6, pp. 6603–6618, 2020.

10. M. Lee, J. Lee, H. Lee, T. Kim, and S. Pack, "On performance of deep reinforcement learning-based listen-before-talk (LBT) scheme," in *International Conference on Information Networking (ICOIN)*, pp. 72–75, IEEE, 2021.

11. 3GPP.org, "Feasibility study on licensed-assisted access to unlicensed spectrum," release 13, TR 36.889, Sept. 2016.

12. 3GPP.org, "Evolved universal terrestrial radio access (E-UTRA) physical layer procedures," document TS 36.213, v13.6.0, June 2017.

13. M. Alhulayil and M. Lopez-Benitez, "Novel LAA waiting and transmission time configuration methods for improved LTE-LAA/Wi-Fi coexistence over unlicensed bands," *IEEE Access*, vol. 8, pp. 162373–162393, 2020.

14. V. Sathya, M. I. Rochman, and M. Ghosh, "Measurement-based coexistence studies of LAA & Wi-Fi deployments in Chicago," *IEEE Wireless Communications*, vol. 28, no. 1, pp. 136–143, 2020.

15. S. Saadat, W. Ejaz, S. Hassan, I. Bari, and T. Hussain, "Enhanced network sensitive access control scheme for LTE–LAA/WiFi coexistence: Modeling and performance analysis," *Computer Communications*, vol. 172, pp. 45–53, 2021.

16. S. Saadat, D. Chen, K. Luo, M. Feng, and T. Jiang, "License assisted access-WiFi coexistence with TXOP backoff for LTE in unlicensed band," *China Communications*, vol. 14, no. 3, pp. 1–14, 2017.

17. G. Bianchi, "Performance analysis of the IEEE 802.11 distributed coordination function," *IEEE Journal on Selected Areas in Communications*, vol. 18, pp. 535–547, Mar. 2000.

18. Y. Song, K. W. Sung, and Y. Han, "Coexistence of wi-fi and cellular with listen-before-talk in unlicensed spectrum," *IEEE Communications Letters*, vol. 20, pp. 161–164, Jan. 2016.

19. M. Yazid, A. Ksentini, L. Bouallouche-Medjkoune, and D. Aissani, "Performance analysis of the TXOP sharing mechanism in the VHT IEEE 802.11 ac WLANs," *IEEE Communications Letters*, vol. 18, no. 9, pp. 1599–1602, 2014.

20. S. Mosleh, Y. Ma, J. D. Rezac, and J. B. Coder, "A novel machine learning approach to estimating KPI and PoC for LTE-LAA-based spectrum sharing," in *IEEE International Conference on Communications Workshops (ICC Workshops)*, pp. 1–6, IEEE, 2020.

21. A. Sahoo, "A machine learning based scheme for dynamic spectrum access," in *IEEE Wireless Communications and Networking Conference (WCNC)*, pp. 1–7, IEEE, 2021.

22. S. Liu, L. Xiao, Z. Han, and Y. Tang, "Eliminating NB-IoT interference to LTE system: A sparse machine learning-based approach," *IEEE Internet of Things Journal*, vol. 6, no. 4, pp. 6919–6932, 2019.

23. D. Magrin, C. Pielli, C. Stefanovic, and M. Zorzi, "Enabling LTE RACH collision multiplicity detection via machine learning," in *International Symposium on Modeling and Optimization in Mobile, Ad Hoc, and Wireless Networks (WiOPT)*, pp. 1–8, IEEE, 2019.

24. A. Dziedzic, V. Sathya, M. I. Rochman, M. Ghosh, and S. Krishnan, "Machine learning enabled spectrum sharing in dense LTE-U/Wi-Fi coexistence scenarios," *IEEE Open Journal of Vehicular Technology*, vol. 1, pp. 173–189, 2020.

25. M. Girmay, A. Shahid, V. Maglogiannis, D. Naudts, and I. Moerman, "Machine learning enabled Wi-Fi saturation sensing for fair coexistence in unlicensed spectrum," *IEEE Access*, vol. 9, pp. 42959–42974, 2021.

26. S. Aoki, K. Shiomoto, and C. L. Eng, "Few-shot learning and self-training for eNodeB log analysis for service-level assurance in LTE networks," *IEEE Transactions on Network and Service Management*, vol. 17, no. 4, pp. 2077–2089, 2020.

27. J. Tan, L. Zhang, Y.-C. Liang, and D. Niyato, "Deep reinforcement learning for the coexistence of laa-lte and wifi systems," in *International Conference on Communications (ICC)*, pp. 1–6, IEEE, 2019.

28. A.-T. H. Bui and A. T. Pham, "Deep reinforcement learning-based access class barring for energy-efficient MMTC random access in LTE networks," *IEEE Access*, vol. 8, pp. 227657–227666, 2020.

29. A. Robinson and T. Kunz, "Downlink scheduling in LTE with deep reinforcement learning, LSTMs and pointers," in *MILCOM 2021–2021 IEEE Military Communications Conference (MILCOM)*, pp. 763–770, IEEE, 2021.

30. N. Stepanov, D. Alekseeva, A. Ometov, and E. S. Lohan, "Applying machine learning to LTE traffic prediction: Comparison of bagging, random forest, and SVM," in *12th International Congress on Ultra Modern Telecommunications and Control Systems and Workshops (ICUMT)*, pp. 119–123, IEEE, 2020.

31. B. Zeng, Y. Zhong, and X. Niu, "A factorization machine-based approach to predict performance under different parameters in cellular networks," *IEEE Access*, vol. 8, pp. 113142–113150, 2020.

32. J. M. Sierra, "Wifi data." https://www.kaggle.com. Accessed on Feb. 24, 2022.

10 Programming Languages, Tools, and Techniques

Muhammad Ateeq

Department of Data Science, Faculty of Computing,
The Islamia University of Bahawalpur, Pakistan

Muhammad Khalil Afzal

COMSATS University Islamabad, Wah Campus, Pakistan

CONTENTS

DOI: 10.1201/9781003216971-13

10.1 INTRODUCTION

With the advent of data-driven techniques, it became inevitable that practical support would be necessary to facilitate these beneficial techniques. Because the systems are largely controlled through software, programming languages are the primary focus to provide necessary constructs and tool kits to adequately implement the data-driven models [1]. In addition, the development of custom tools that can automate the known processes and hide the implementation from users is considered productive and promotive. Such tools can help primitive users by easing implementation and can save time for advanced users by allowing more focus on the problem at hand.

An important aspect of realizing a data-driven network is to devise practical frameworks and process flows that can be followed in order to benefit from this promising artifact [2, 3]. In theory, the methods for handling and learning from data have been proposed and explored at length. These studies and findings greatly benefit the proponents of data-driven techniques in the communication discipline. These methods include procedures such as cleaning and completing data, analyzing and visualizing the data, identifying and posing the right questions, and learning from data.

This chapter is divided into five sections; in Section 10.2 we survey the popular programming languages such as Python and R with relevant libraries (e.g., pandas, NumPy, Matplotlib, Scikit-learn, data.table, dplyer, and ggplot2). Section 10.3 comprises useful tools such as Weka, Orange, and RapidMiner. The flow of an example generic process is presented in Section 10.4 and the conclusion and challenges are listed in Section 10.5.

10.2 PROGRAMMING LANGUAGES

In general, data science finds growing support in programming languages. Python [4] and R [5] are considered among the most prominent examples in this context. In the following, we describe both programming languages and cover their relevant features and libraries.

10.2.1 PYTHON

Python, with its generalized programming language design, ease of learning, and wide applicability, is considered the most obvious choice when applying data-driven techniques and models in any domain. The pseudocode-like syntax makes it easy to learn the language and develop solutions, allowing the user to better focus on the problem rather than language learning.

Data-driven techniques require support for descriptive and inferential data analysis, data visualization, data cleaning, data transformations, statistical measures, and machine learning among others. Python with its rich support of libraries and an open-source community serves the needs adequately. Next, we give a brief overview of the libraries related to data science [6].

10.2.1.1   NumPy

NumPy is short for Numerical Python and provides the foundation for numerical operations. NumPy is among the most actively maintained and contributed libraries of Python. It implements multi-dimensional arrays that are expressive and fast. To meet the speed challenges, NumPy provides a wide range of built-in functions that are optimized to do well with the NumPy arrays.

Some of the prominent features of NumPy are:

- Provides fast, one and multi-dimensional array objects.
- Facilitates interfacing with code from other languages like C/C++ and Fortran.
- Supports vectorization for fast numerical computations.
- Provides a solid foundation for other libraries like Pandas, Scikit-learn, and SciPy (scientific Python).

10.2.1.2   Pandas

Pandas is generally the foremost library a data scientist should learn along with NumPy and Matplotlib. Pandas is amongst the most actively maintained Python libraries. It is used for all preparatory and initial analytical steps involved in data cleaning and analysis. In addition, it supports some primitive visualizations. To support wide operations on large data sets, Pandas uses NumPy at the backend and is usually blazingly fast compared to conventional programming constructs provided by Python language.

Some of the prominent features of Pandas are:

- Complete documentation that explains the whole library, with examples.
- Expressive set of functions and features that enable adequate data handling.
- Provides API for custom development and contribution to the library.
- Implements complex and demanding data operations with a high level of abstraction.
- Supports operations like data wrangling and data cleaning adequately.

10.2.1.3   Matplotlib

Matplotlib is the primary library for visualization and plotting provided for Python. Due to wide support and contributions, Matplotlib is considered a competitive library when compared to any other platform, programming language, or tool. An object-oriented API makes sure that the visualizations created with Matplotlib can be integrated into other applications with ease.

Some of the prominent features of Matplotlib are:

- An open-source and free competitor to MATLAB with adequate features and facilities.
- Implements diverse output types that are useable across different platforms without any special assistance.
- Efficient in using system resources, thus, plots from big data sets can be created with sufficient ease.
- Integrates necessary statistical analysis and measures like correlations, confidence intervals, etc.
- Behaves as a base for a sophisticated library like Seaborn.

10.2.1.4 Scikit-learn

Scikit-learn is Python's prolific machine learning library. It implements almost all machine learning models and relevant concepts and works well with Pandas, NumPy, and other relevant libraries. It supports both supervised and unsupervised learning.

Some of the prominent features of Scikit-learn are:

- Supports pre-processing including transformation, normalization, and encoding.
- Provides the implementation of classification and regression models based on linear, non-linear, gradient descent, tree-based, Bayesian, and ensemble methods.
- Supports clustering for unsupervised learning implementing methods such as K-mean, affinity propagation, spectral, and hierarchical among others.
- Implements dimensionality reduction covering principal component analysis, independent component analysis, and latent Dirichlet allocation.

Some other important libraries that turn out to be important at various stages of data-driven systems include SciPy, TensorFlow Keras and PyTorch (for deep learning), Scrapy, and BeautifulSoup (for scrapping and dealing with semi-structured or unstructured data).

10.2.2 R

R is a programming language designed for statistical computing. Like Python, it is interpreted as well as open source. However, unlike Python, R is more confined and specialized for data processing and does not find appreciation in solving computing problems. The language supports data types like lists, vectors, arrays, and data frames that make it very convenient to process data.

As expected, R has a wide range of packages related to data manipulation and processing. Next, we briefly explain some of the prominent packages and libraries by category [7].

10.2.2.1 Data Loading

R offers broad support in terms of loading data from diverse formats and sources. R can be used to read plain data files without needing any package. To read data from databases, packages like DBI and ODBC are available. XLConnet and xlsx are the packages that read and write Excel files as CSVs. R also interfaces with software like Statistical Product and Service Solution (SPSS) or Stata, using packages like foreign and haven.

10.2.2.2 Data Manipulation and Visualization

A collection of relevant packages is available in the form of tidyverse. The primary package for data manipulation is dplyr. It provides data manipulation similar to Pandas and is quite fast. In addition, there are useful packages for data manipulation like tidyr, stringr, and lubridate.

R also has several packages for data visualization. Ggplot2 is the main library in R to create feature-rich custom visuals. In addition, ggviz, rgl, htmlwidgets, and googleViz are all relevant tools for visualization in R.

10.2.2.3 Data Modeling

Statistical as well as machine learning model implementation is available in R through several packages. Tidymodels is a comprehensive collection of packages that provide popular machine learning models for R programmers. In addition, there are a lot of specialized packages for various machine learning and statistical models including car (having Anova functions), mgcv (additive models), multcomp, vcd, glmnet, and caret.

10.2.2.4 Other Packages

In addition to the packages discussed previously, R has great support for data scientists through additional packages for various tasks. For example, it provides shiny, xtable, and Markdown for reporting results. Sp, maptools, maps, and ggmap are popular when dealing with spatial data. Zoo, xts, and quantmod are useful for analyzing time series and financial data. R also provides the ability to write your own packages through devtools, testthat, and roxygen2.

10.3 TOOLS

In this section, we list a few important tools useful in adopting and implementing data-driven methods. In this context, we discuss three different categories of tools. In the first category, we place the tools that are capable of handling large-scale data. The second category belongs to the tools useful for data analysis, and the third category discusses the tools for predictive modeling and machine learning.

10.3.1 BIG DATA

Intuitively, data-driven techniques are driven by some form of data at the backend. Data can be structured, semi-structured, or even unstructured. They can be numbers, text, images, or any combination of these. When it comes to dealing with communication and networks, the data can be some performance statistics coming from traffic logs or signals. In general, any kind of statistics measured from traffic logs present data in a structured way. Even if the data are representing some kind of signals, they are often represented in the form of numerical quantities. Therefore, communication and networks often deal with structured data.

Normal-sized data can be represented using text files where CSV is a popular format. However, larger data sets require better support to store, retrieve, and process the data at scale. Next, we briefly discuss some important and useful tools for big data handling [8, 9].

10.3.1.1 Hadoop

Hadoop offers solutions to solve big data problems using a network of computers. It is a collection of open-source libraries with the Hadoop distributed file system at the core for data storage and uses MapReduce as a programming model. It uses YARN to manage computing resources in the clusters and schedules applications. However, a major limitation is that MapReduce can run one job at a time in batch processing mode. This limits the usefulness of Hadoop as a real-time analysis framework and makes it a prominent choice for data warehousing.

10.3.1.2 Spark

Spark is popular for real-time data-stream processing in the context of big data systems. Although the workflows used in Spark are based on Hadoop MapReduce, Spark's are more efficient because it provides its own streaming API rather than banking on Hadoop YARN. This makes Spark more suitable for real-time data-stream processing than Hadoop, which has turned out to be better as a tool for storage and batch processing.

Spark banks on data stored in Hadoop and does not implement its own storage system. For the sake of development, Spark uses Scala tuples.

10.3.1.3 Cloudera

Cloudera was developed as an enterprise-level deployment solution based on Hadoop. It can interact and access data from heterogeneous environments, offering real-time analysis. Cloudera can interact with different clouds, thus, implementing truly enterprise-wide solutions. In addition to data analysis, it also provides the capability to train and deploy data models. Cloudera is versatile in that it can be deployed across multiple clouds as well as on-site and is a popular choice to implement business intelligence solutions.

Cloudera provides multiple language support options for application development including C/C++, Python, Scala, Go, and Java.

10.3.1.4 MongoDB

MongoDB is a free solution for implementing databases that can overcome the limitations of relational databases and is based on a NoSQL design scheme. MongoDB can handle a large amount of data beyond the traditional structure followed by relational databases. MongoDB Atlas enables developers to manage databases across different cloud providers including Azure, AWS, and Google Cloud. It supports more than 10 languages.

Some of the prominent features of MongoDB are listed here. The game-changing feature of MongoDB is that it supports real-time analytics based on ad hoc queries.

Moreover, its indexing and data replication features demonstrate great performance advantages. The load balancing feature also outperforms many competing solutions.

10.3.2 DATA ANALYSIS

There are a lot of tools that provide users with the capability to do primitive and advanced statistical analysis of data without having to program anything. In general, tools have built-in methods to carry out the implemented tasks by following a well-defined sequence of steps. This is why tools are easy for primitive users and useful for advanced users in carrying out some initial analysis or handling simplistic situations without having to program anything.

Although there are numerous tools available for data analysis, next we discuss two broad tools [10].

10.3.2.1 Spreadsheets

Spreadsheets are the most widely used software to handle structured data arranged into rows and columns. Amongst the most-used spreadsheet software titles are Microsoft Excel and Google Sheets. Excel is a proprietary package whereas Google Sheets are free and also available through Google Cloud in the form of software-as-a-service. Numerical operations, statistical formulas, custom functions, data handling, and basic plotting are some of the prominent and most-used features in spreadsheets.

10.3.2.2 SPSS

SPSS is a proprietary solution by IBM for advanced statistical analysis. It has been widely used in the social science domains as well as by market researchers, healthcare and survey companies, education, and various government sector setups.

The noticeable features of SPSS include descriptive statistical analysis, statistical tests, simple predictive modeling, text analysis, and visualizations.

10.3.3 MACHINE LEARNING

Although the libraries for machine learning are well implemented and available in programming languages like Python and R, there are certain tools in the market that make it convenient for users to be able to do predictive modeling without having to program explicitly. Here, we briefly introduce Weka and RapidMiner, two of the most widely used tools for machine learning and data mining [11].

10.3.3.1 Weka

Weka is free software developed at the University of Waikato for the purpose of data analysis and predictive modeling. It provides support for both supervised and unsupervised machine learning. The features implemented in Weka include pre-processing, classification, regression, and clustering.

Data in Weka can be read from files of different formats, Web via URL, as well as from databases. Weka makes it easy to see the behavior of various machine learning models on the data sets of interest without having to have explicit programming knowledge.

10.3.3.2   Orange

Orange is a cross-platform free software for data analysis and machine learning. Like Weka, it implements all basic machine learning models. In addition to the features provided by Weka, Orange is Python-based and includes support for plug-ins. It also implements support for text processing and simulations.

Both Weka and Orange are useful software for testing machine learning models with ease. The support is adequate and learning is rather easy. In addition, there are several other software packages like RapidMiner, KNIME, Neural Designer, and KEEL.

10.4   AN EXAMPLE FRAMEWORK

In this section, we sketch the flow of the overall process for data-driven solutions in wireless networks. Although the presentation is generic, to make concepts and process flow more concrete, wireless sensor networks (WSNs) and Internet of Things (IoT) are referred to, more specifically. We argue that a system can achieve an adaptive design based on the parameters found within the system by dynamically understanding the relationship between those parameters (e.g., transmission power, traffic rate, backoff window, etc.) and various performance indicators (e.g., signal strength, delay, packet delivery ratio, etc.). A pictorial representation is shown in Figure 10.1. It can be observed that information from various deployments of heterogeneous networks consisting of different parameters of the protocol stack of the relevant network can be processed by the intelligent servers to identify the relation between observed values of parameters and quality of service (QoS) metrics. This can help achieve a predictive and adaptive design.

Intuitively, any data-driven system first conceives the source of data and how it is collected. The data collected may go through a process of cleaning, completion, transformation, etc. Analyzing and visualizing the data is the next general step.

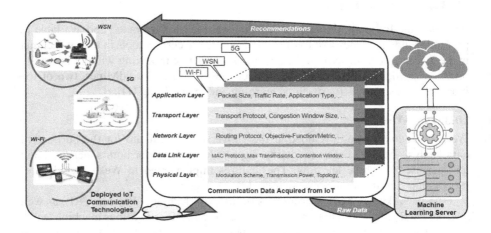

FIGURE 10.1   Identified parameters to bring cognition in quality-of-service predictions.

Depending on the problem, machine learning and predictions can be an important part of it. Finally, it matters how the results are made part of the system in production.

Here, we describe the important stages of a data-driven solution designed for wireless networks.

10.4.1 DATA COLLECTION

As with all systems based on data-driven design, data has pivotal significance in achieving an adaptive and self-tuning solution for any aspect of wireless networks. Generally, the data collected from real systems is often processed offline and recommendations are prepared discretely for predictive measures. However, considering any real-world wireless system, data must be collected and processed interactively, and real-time decisions are to be communicated to the real-world deployments in order to make real-time decisions. Considering the limitations, and from a design-in-research perspective, data can be collected from the public data repositories, testbeds, or generated using simulations.

Here, we analyze the potential data sources [12] and explain their benefits and drawbacks.

10.4.1.1 Simulators

Historically, it has not been possible to realize real deployments of desired network technologies and topologies for research purposes. Therefore, there are several simulators available and in use to create design scenarios and evaluate the performance of wireless networks.

Some popular examples include ns-2, ns-3, OMNET++, and COOJA [12]. Where it has been possible to create custom network topologies with desired software and hardware characteristics and configurations, simulations fail to facilitate the behavior of real deployments because even the stochastic events are deterministic.

10.4.1.2 Testbeds

The second possibility is to generate the data using a testbed. With the proliferation of wireless communication, IoT is very much a realization on the timeline. This has induced a lot of interest from the research and commercial and governmental organizations to create suitable testbeds to foster research in the domain of wireless communications. Some prominent examples of accessible testbeds include MoteLab, TWIST, and Indriya [12]. Lately, FED4FIRE+ has federated a large set of testbeds, focusing on diverse networking and cloud facilities.

10.4.1.3 Public Sources

Public data sources are of paramount importance for research driven by data. Although there are some wireless data sets related to QoS performance, these do not comprehend the diverse deployment scenarios and application requirements and also do not cope with the evolving nature of network design. Most of the data set primarily focuses on the sensed information rather than communication performance. A prominent public data set providing comprehensive measurements based on a large combination of diverse parameter settings in WSNs [13] is hosted by CRAWDAD, a large public repository for networking-related data sets.

10.4.1.4 Real-World Deployments

Real-time decision-making is critical, particularly for time-critical safety-related scenarios. The purpose of adopting data-driven QoS prediction is to facilitate real-world deployments of communication systems forming the IoT. These real-world deployments have their own dynamic and evolving nature. Therefore, it is more desirable and effective to acquire data from these real scenarios. Moreover, this evolution can only be accommodated through the integration of prediction systems with real deployments. However, gathering data from live sources and carrying out real-time analytics for QoS still requires considerable attention.

10.4.2 DATA ANALYSIS AND MACHINE LEARNING

The data collected from desired network deployments require pre-processing and is generally shaped in the form of features (which could be the configurable parameters from various layers of the protocol stack) and targets (which could be various QoS metrics, security labels, etc.). A sample of what the feature vector could look like is shown in Figure 10.2. The prediction of QoS metrics involves identifying the correct set of features for a particular metric, and feature selection techniques can help. Big data platforms and services like Hadoop, Kafka, MQTT, Spark Streaming, etc., play a vital role in putting the system components together. FED4FIRE+ facilitates testbeds like w-iLabt.t for experiments involving WSNs, Wi-Fi, LTE/5G, cognitive

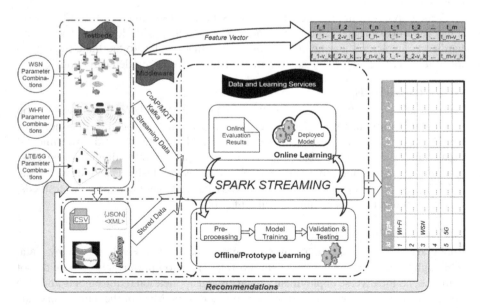

FIGURE 10.2 Architecture and flow of events in data-driven the quality-of-service framework.

radio, etc. The big data platform like Tengu is available with adequate facilities for hosting big data and provides streaming services for live interactions. To provide gluing between the testbeds and big data platforms, seamless connectivity is delivered by a virtual wall. The next important thing is choosing a suitable machine learning model(s) that can meet the desired performance requirements at an affordable cost. As an example, we have used deep neural networks for predicting QoS in WSNs [14–16]. Using real-time analytics for configuring the nodes and network can serve the real advantage of achieving adaptivity and self-reconfigurability.

10.4.3 Dissemination of Recommendations

Recommendations are formulated consisting of suitable values for critical features considering the QoS metrics and are disseminated to the interested nodes in the networks. This way, for each threshold of a prediction target, a recommended set of values for the critical set of features can be provided proactively or reactively to the sensor nodes. The sole task that the wireless nodes need to perform is to choose the right set of values by considering a performance goal that is computationally very simple.

REFERENCES

1. Ozgur, Ceyhun, Taylor Colliau, Grace Rogers, and Zachariah Hughes, "MatLab vs. Python vs. R," Journal of Data Science 15, no. 3 (2017): 355–371.
2. Dhar, Vasantm, "Data science and prediction," Communications of the ACM 56, no. 12 (2013): 64–73.
3. Provost, Foster and Tom Fawcett, "Data science and its relationship to big data and data-driven decision making," Big Data 1, no. 1 (2013): 51–59.
4. Python, https://www.python.org/, accessed July 12, 2022.
5. R, https://www.r-project.org/, accessed July 12, 2022.
6. Stančin, Igor and Alan Jović, "An overview and comparison of free Python libraries for data mining and big data analysis," In 42nd International Convention on Information and Communication Technology, Electronics and Microelectronics (MIPRO), pp. 977–982. IEEE, 2019.
7. Top R libraries for Data Science, https://towardsdatascience.com/top-r-libraries-for-data-science-9b24f658e243, accessed July, 12, 2022.
8. Arfat, Yasir, Sardar Usman, Rashid Mehmood, and Iyad Katib, "Big data tools, technologies, and applications: A survey," In Smart Infrastructure and Applications, pp. 453–490. Springer Cham, 2020.
9. Ratra, Ritu, and Preeti Gulia, "Big data tools and techniques: A roadmap for predictive analytics," International Journal of Engineering and Advanced Technology (IJEAT) 9, no. 2 (2019): 4986–4992.
10. Cleff, Thomas, Applied Statistics and Multivariate Data Analysis for Business and Economics: A Modern Approach Using SPSS, Stata, and Excel. Springer, 2019.
11. Naik, Amrita and Lilavati Samant, "Correlation review of classification algorithm using data mining tool: Weka, Rapidminer, Tanagra, Orange and KNIME," Procedia Computer Science 85 (2016): 662–668.
12. Papadopoulos, Georgios Z., Kosmas Kritsis, Antoine Gallais, Periklis Chatzimisios, and Thomas Noel, "Performance evaluation methods in ad hoc and wireless sensor networks: A literature study," IEEE Communications Magazine 54, no. 1 (2016): 122–128.

13. Songwei Fu and Yan Zhang. CRAWDAD dataset due/packet-delivery (v.2015-04-01). Downloaded from https://crawdad.org/due/packet-delivery/20150401, April 2015.

14. Ateeq, Muhammad, Farruh Ishmanov, Muhammad Khalil Afzal, and Muhammad Naeem, "Multi-parametric analysis of reliability and energy consumption in IoT: A deep learning approach," Sensors 19 (2019): 2,309.

15. Ateeq, Muhammad and Farruh Ishmanov, Muhammad Khalil Afzal, and Muhammad Naeem, "Predicting delay in IoT using deep learning: A multiparametric approach," IEEE Access 7 (2019): 62022–62031.

16. Ateeq, Muhammad, Muhammad Khalil Afzal, Muhammad Naeem, Muhammad Shafiq, and Jin-Ghoo Choi, "Deep learning-based multiparametric predictions for IoT," Sustainability 12, no. 18 (2020): 7752.

Index